ANATOMY &
100 ESSENTIAL
STRETCHING
EXERCISES

BARRON'S

ANATOMY &
100 ESSENTIAL
STRETCHING
EXERCISES

Published by Kaplan, Inc., d/b/a Barron's Educational Series
750 Third Avenue
New York, NY 10017
www.barronseduc.com

ISBN: 978-1-4380-0617-8

Library of Congress Control No.: 2014940039

Editorial Director: Maria Fernanda Canal
Editing: Ángeles Tomé
Text: Guillermo Seijas Albir
Technical Review: Ana Lorenzo
Typographical Correction: Roser Pérez
Graphic Design: Toni Inglès
Illustrations: Myriam Ferrón
Photographs: Nos I Soto
Layout: Estudi Toni Inglès

Printed in Spain
10 9 8 7 6

First edition for the United States,
its territories and dependencies,
and Canada published in 2015
by Kaplan, Inc.,
d/b/a Barron's Educational Series

English-language translation © copyright
2014 by Kaplan, Inc.,
d/b/a Barron's Educational Series
English translation by Eric A. Bye, M.A.

Original Spanish title: *Anatomía & 100
Estiramientos Esenciales*
© Copyright 2015 Editorial
Paidotribo—World Rights
Published by Editorial Paidotribo,
Badalona, Spain

Kaplan, Inc., d/b/a Barron's Educational
Series print books are available at
special quantity discounts to use for
sales promotions, employee premiums,
or educational purposes. For more
information or to purchase books, please
call the Simon & Schuster special sales
department at 866-506-1949.

Preface

Flexibility may be the most overlooked of the basic physical characteristics, and yet it is indispensable for improving performance in any sporting discipline, whether involving strength, resistance, or speed. We take note of individuals who can flip over a tractor tire or run a marathon, and yet we don't think it extraordinary that some people can place their palms on the floor while keeping their knees straight.

Flexibility is a quality that recedes prematurely, and we generally do not realize how quickly we lose it until a lack of mobility becomes so limiting that we cannot scratch our backs, we feel discomfort in turning our heads, or it becomes difficult for us to tie our shoes.

In addition, most muscle pains that people experience, especially if they don't do regular physical activity, are due to muscle imbalances. These imbalances could be rectified easily by following a proper stretching program that would take only a few minutes per day and could be done at home, at the workplace, or practically anywhere, because most stretching exercises require little or no equipment.

The excess number of hours that we spend at the computer, with bad posture, with a misaligned spinal column, or on intensive practice of a specific sport—all of these are factors that cause problems for us sooner or later.

Whether you are thinking about your health or searching for ways to improve your athletic performance, this book will show you the basic pillars that support training for flexibility: how to stretch, what techniques to use, how long to hold a stretch, and how many repetitions to perform.

This information is presented clearly and concisely, and in a way that is detailed and easy to understand, even if you are a beginner in training for flexibility. We intend this to be a complete tool that provides a profound analysis of developing flexibility and simultaneously serves as a quick reference work for any user, from beginners to those who already have some knowledge of the subject matter or are at an advanced level.

Whatever your goal or your starting level, we hope that the information assembled in this book will help you along the road to improving your physical performance and your well-being.

Contents

How to Use This Book

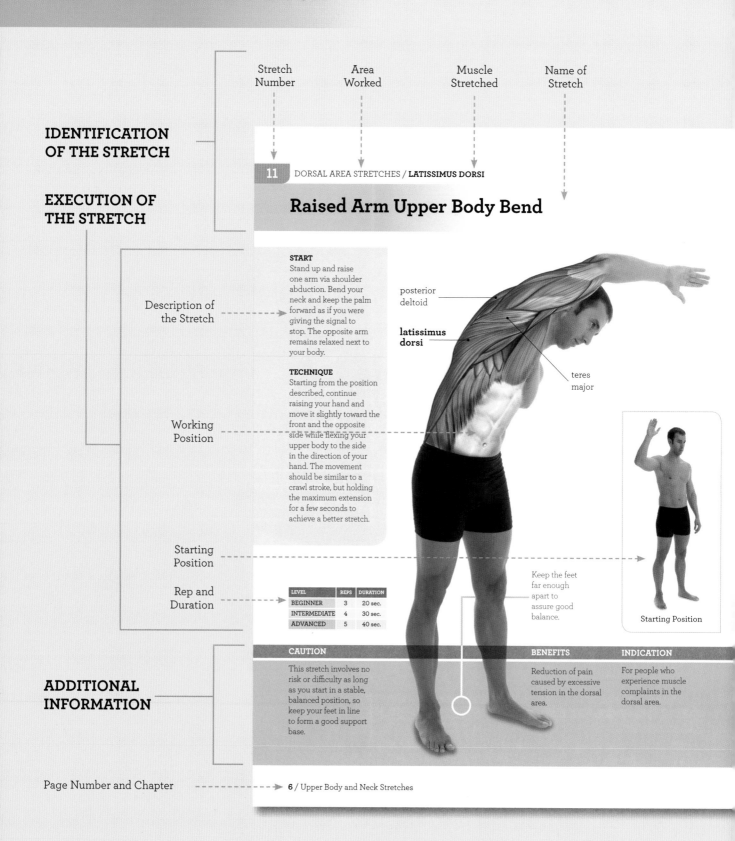

Stretch Number | Area Worked | Muscle Stretched | Name of Stretch

IDENTIFICATION OF THE STRETCH

EXECUTION OF THE STRETCH

11 DORSAL AREA STRETCHES / **LATISSIMUS DORSI**

Raised Arm Upper Body Bend

Description of the Stretch

START
Stand up and raise one arm via shoulder abduction. Bend your neck and keep the palm forward as if you were giving the signal to stop. The opposite arm remains relaxed next to your body.

TECHNIQUE
Starting from the position described, continue raising your hand and move it slightly toward the front and the opposite side while flexing your upper body to the side in the direction of your hand. The movement should be similar to a crawl stroke, but holding the maximum extension for a few seconds to achieve a better stretch.

Working Position

Starting Position

Rep and Duration

posterior deltoid

latissimus dorsi

teres major

LEVEL	REPS	DURATION
BEGINNER	3	20 sec.
INTERMEDIATE	4	30 sec.
ADVANCED	5	40 sec.

Keep the feet far enough apart to assure good balance.

Starting Position

ADDITIONAL INFORMATION

CAUTION

This stretch involves no risk or difficulty as long as you start in a stable, balanced position, so keep your feet in line to form a good support base.

BENEFITS

Reduction of pain caused by excessive tension in the dorsal area.

INDICATION

For people who experience muscle complaints in the dorsal area.

Page Number and Chapter

6 / Upper Body and Neck Stretches

One stretch per page,
like an index card.

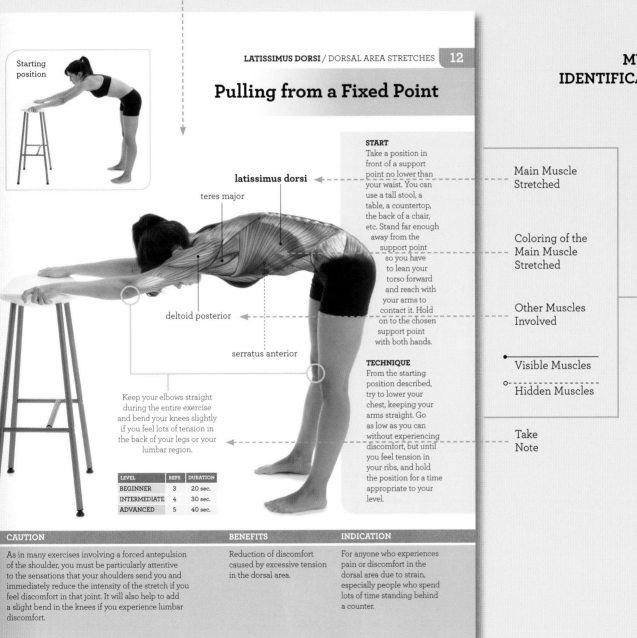

Starting
position

Pulling from a Fixed Point

latissimus dorsi

teres major

deltoid posterior

serratus anterior

Keep your elbows straight
during the entire exercise
and bend your knees slightly
if you feel lots of tension in
the back of your legs or your
lumbar region.

LEVEL	REPS	DURATION
BEGINNER	3	20 sec.
INTERMEDIATE	4	30 sec.
ADVANCED	5	40 sec.

START
Take a position in
front of a support
point no lower than
your waist. You can
use a tall stool, a
table, a countertop,
the back of a chair,
etc. Stand far enough
away from the
support point
so you have
to lean your
torso forward
and reach with
your arms to
contact it. Hold
on to the chosen
support point
with both hands.

TECHNIQUE
From the starting
position described,
try to lower your
chest, keeping your
arms straight. Go
as low as you can
without experiencing
discomfort, but until
you feel tension in
your ribs, and hold
the position for a time
appropriate to your
level.

CAUTION

As in many exercises involving a forced antepulsion
of the shoulder, you must be particularly attentive
to the sensations that your shoulders send you and
immediately reduce the intensity of the stretch if you
feel discomfort in that joint. It will also help to add
a slight bend in the knees if you experience lumbar
discomfort.

BENEFITS

Reduction of discomfort
caused by excessive tension
in the dorsal area.

INDICATION

For anyone who experiences
pain or discomfort in the
dorsal area due to strain,
especially people who spend
lots of time standing behind
a counter.

MUSCLE
IDENTIFICATION

Main Muscle
Stretched

Coloring of the
Main Muscle
Stretched

Other Muscles
Involved

Visible Muscles

Hidden Muscles

Take
Note

Anatomical Atlas
Locations of the Muscles

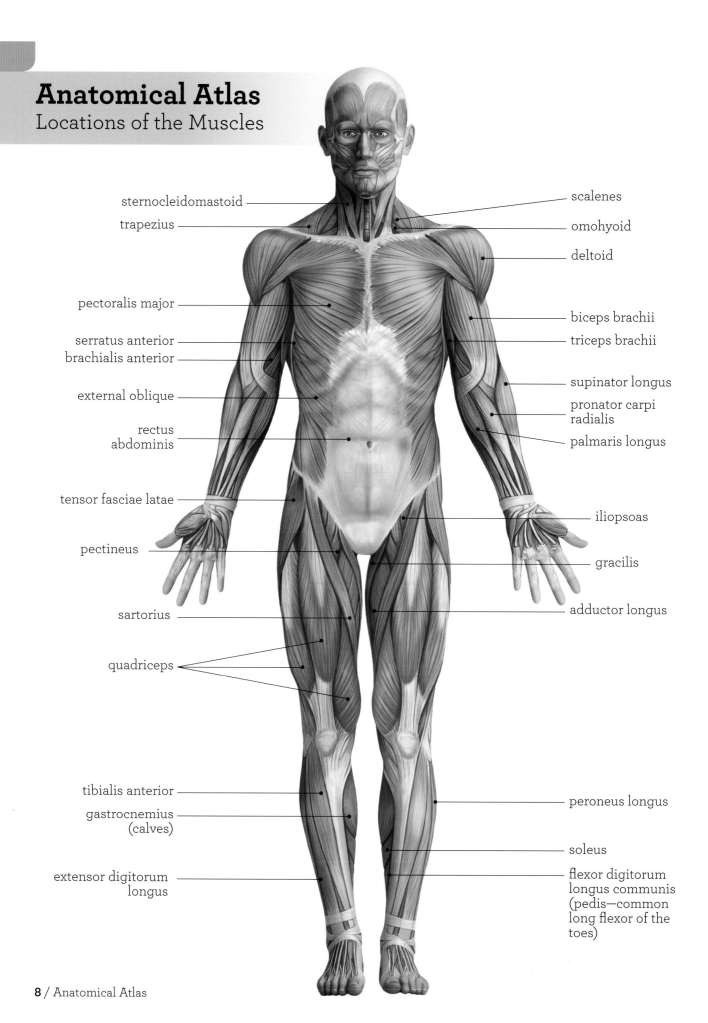

sternocleidomastoid

trapezius

pectoralis major

serratus anterior

brachialis anterior

external oblique

rectus abdominis

tensor fasciae latae

pectineus

sartorius

quadriceps

tibialis anterior

gastrocnemius (calves)

extensor digitorum longus

scalenes

omohyoid

deltoid

biceps brachii

triceps brachii

supinator longus

pronator carpi radialis

palmaris longus

iliopsoas

gracilis

adductor longus

peroneus longus

soleus

flexor digitorum longus communis (pedis—common long flexor of the toes)

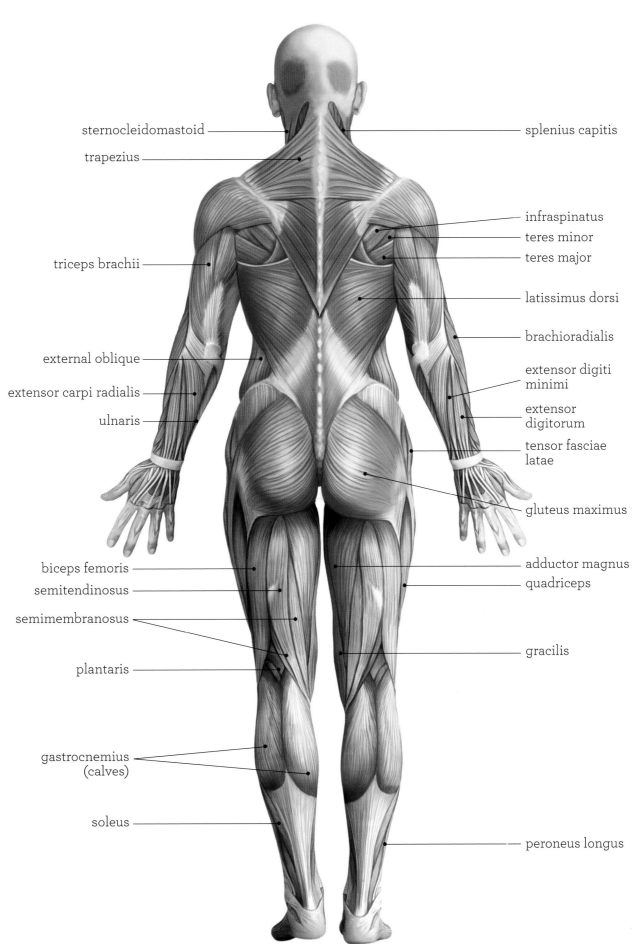

sternocleidomastoid

trapezius

triceps brachii

external oblique

extensor carpi radialis

ulnaris

biceps femoris

semitendinosus

semimembranosus

plantaris

gastrocnemius
(calves)

soleus

splenius capitis

infraspinatus

teres minor

teres major

latissimus dorsi

brachioradialis

extensor digiti
minimi

extensor
digitorum

tensor fasciae
latae

gluteus maximus

adductor magnus

quadriceps

gracilis

peroneus longus

Planes of Movement

Before starting, it is appropriate to explain a series of terms that refer to body movements and will appear consistently throughout the book. If you do not know the basic nomenclature of the movements, it will be difficult to understand the detailed description of the exercises. Some of these terms, such as *flexion* and *extension*, are commonly used; but, others, such as *inversion, eversion, adduction,* and *supination* are used in narrower circles, so it can be very helpful to clarify their meaning.

The first thing we need to know is that body movements occur in three different planes: the frontal, the sagittal, and the transverse. Each of these planes involves a certain series of movements, as we shall see below. To understand them, we can begin with the basic anatomical position, which is shown in the illustration.

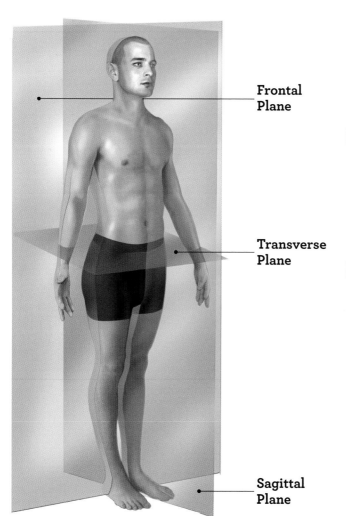

Frontal Plane

Transverse Plane

Sagittal Plane

ABDUCTION

ADDUCTION

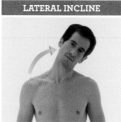
LATERAL INCLINE

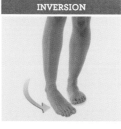
INVERSION

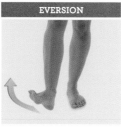
EVERSION

FRONTAL PLANE

This plane divides the body into ventral and dorsal parts—in other words, anterior and posterior. The chest and stomach are in the ventral part, and the back of the neck, the back, and the buttocks are in the dorsal part. The movements of the frontal plane are the following:

Abduction: This is the movement by which we extend a limb away form the central axis of the body. It is easily seen from in front or the rear of the individual because the change in silhouette is obvious from that vantage point. In forming the shape of a cross with our arms, we perform abduction of the shoulders.

Adduction: This is a movement by which we retract a limb toward the central axis of the body—in other words, the movement opposite abduction. If our arms form a cross and we lower them so they are next to our bodies, we perform an adduction of the shoulders.

Lateral Incline: In this movement the head, the neck, or the upper body is tilted to one side. If we fall asleep while sitting up, generally our heads and necks eventually tip to one side through lateral inclination.

Inversion: Although this movement does not involve the frontal plane exclusively, it is the movement where the frontal plane is most common. Inversion of the foot results when the tip and the sole are turned inward at the same time plantar flexion is performed.

Eversion: This is the movement in which the tip and the sole of the foot are turned outward at the same time that the foot experiences dorsal flexion.

FLEXION

EXTENSION

SAGITTAL PLANE

This plane divides the body into two halves: the right and the left. The movements in this plane are best perceived from one side of the individual, in a profile view. In this plane, the following movements stand out:

Flexion: This is the movement in which we extend a part of the body forward with respect to the central axis. For example, if we bend our elbows, we move our forearms ahead with respect to the central axis. There are exceptions to this definition, such as flexion of the knee and plantar flexion of the ankle.

Extension: A movement by which we extend a part of the body rearward with respect to the central axis, or by which we align a body part with the axis. For example, if we look at the sky while we are standing, we have to perform an extension of the cervical column. Once again, the knee is an exception.

Antepulsion: This is the equivalent of flexion, but it applies solely to shoulder movement.

Retropulsion: This is equivalent to extension, but it applies solely to shoulder movement.

Dorsal Flexion: A bending movement that is applicable solely to the ankle joint.

Plantar Flexion: This term designates the movement of the ankle that is equivalent to extension.

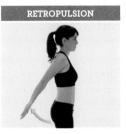

ANTEPULSION

RETROPULSION

DORSAL FLEXION

PLANTAR FLEXION

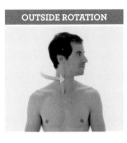

OUTSIDE ROTATION

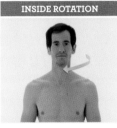

INSIDE ROTATION

PRONATION

SUPINATION

TRANSVERSE PLANE

This plane divides the body into upper and lower parts. Movements in this plane are easily perceived from any angle, but slightly better from the rear or from beneath the person. They include the following:

Outside Rotation: This is the movement in which we turn parts of the body outward and along their own axis. If we are seated at a table and a fellow diner next to us speaks to us, we perform an outside rotation of our necks to look at the person.

Inside Rotation: This movement is the opposite of the preceding one, because it involves turning toward the inside and along the axis of a body part. As we finish the conversation with a fellow diner seated next to us, we perform an inside rotation of our necks to return our gaze toward the front.

Pronation: A rotational movement of the forearm in which we place the back of our hands upward and the palms downward. When we use a knife or a fork to manipulate the food on a plate, our hands are in pronation.

Supination: This is the movement opposite the preceding one—it involves the rotation of the forearm, in which we place the palms facing upward. For example, if someone gives us a handful of seeds, we place our hands palms up, in supination, like a bowl, so we don't drop them.

What Stretching Exercises Are

Physical exercise and staying in good overall shape are the way toward a long, high-quality life. Nowadays, most people are aware of the importance of taking care of their bodies and the positive effect that physical activity has on them.

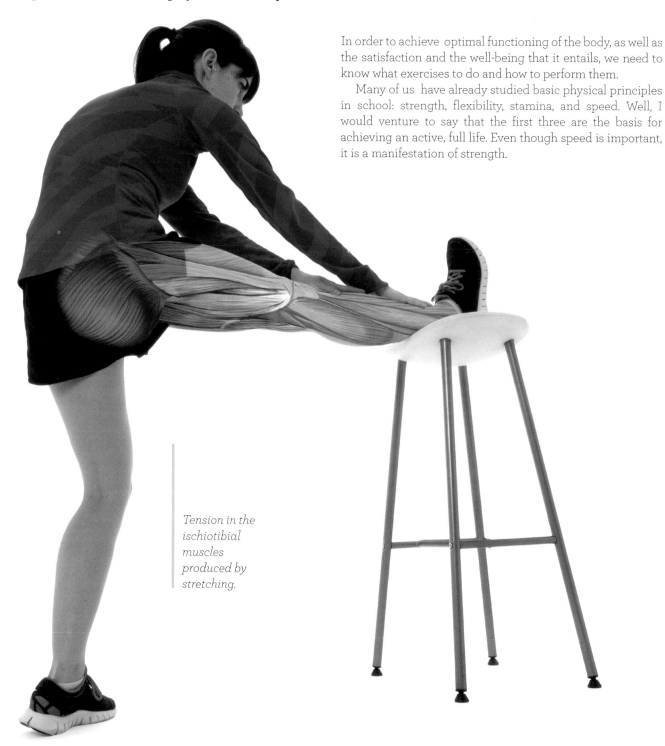

In order to achieve optimal functioning of the body, as well as the satisfaction and the well-being that it entails, we need to know what exercises to do and how to perform them.

Many of us have already studied basic physical principles in school: strength, flexibility, stamina, and speed. Well, I would venture to say that the first three are the basis for achieving an active, full life. Even though speed is important, it is a manifestation of strength.

Tension in the ischiotibial muscles produced by stretching.

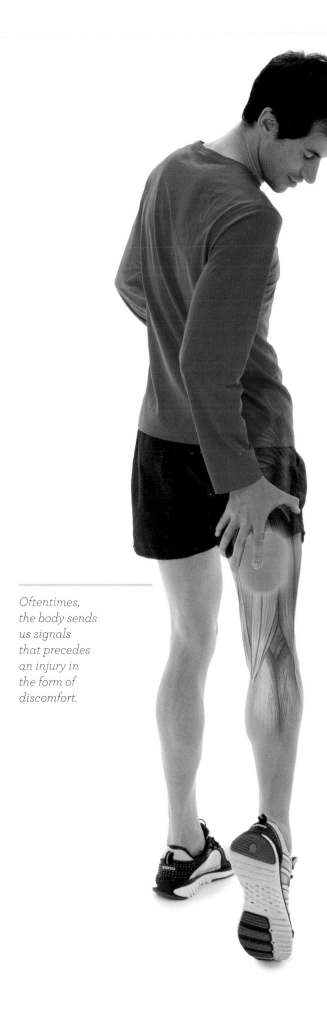

Oftentimes, the body sends us signals that precedes an injury in the form of discomfort.

Training to improve these three qualities will slow down physical deterioration and notably retard the aging process. We all know that there are three very significant physical limitations that appear and become more pronounced with the passage of years:

Loss of Muscle Mass: This occurs most significantly starting at age 35–40. This loss of muscle mass entails a loss of strength; and, as a result, greater difficulty in making a maximum effort.

Loss of Stamina: The passage of time and inactivity cause our cardiovascular system to deteriorate and become increasingly inefficient. So going up a hill or a set of stairs, which some years earlier was no effort for us, now may become an impossible challenge. We literally don't have the wind for it.

Loss of Mobility: Unfortunately, flexibility begins to decrease from the time of our birth. Whereas strength and stamina naturally improve up to adulthood and then begin their decline, flexibility diminishes as soon as we begin to breathe, so we need to take special care with it. Muscle and joint deterioration means that many people reach maturity and old age with serious stiffness and limitation of movement. In addition, we know that a lack of mobility leads to inactivity, and inactivity leads to a greater lack of mobility, and the cycle continues until the individual's passing. But this is not an unavoidable fate. It's up to us whether we resign ourselves to the deterioration or resolve to do something about it, and this is where flexibility training comes in.

Training for flexibility involves stretching: exercises with which we create tension in a muscle to bring it to its maximum length.

Regular stretching will allow us to maintain the optimal range of movement, and thus move about freely, so our bodies will be a wonderful means for relating to our surroundings, rather than a painful and restrictive shell that we cannot shed.

Daily life subjects us to improper posture and imbalanced physical work. Many times, this produces discomfort, pain, fatigue, and a multitude of signals that our bodies send us that we often do not know how to interpret. We are so focused on work or daily tasks that we have forgotten how to enjoy our bodies; and, what's worse, we no longer know how to interpret the sensations that it transmits, the signals it uses to tell us, "I feel fine, I can continue running," or "I am at my limit and can't do any more."

The imbalances that affect untrained individuals may also be present in trained individuals who appear to be in enviable condition. We must remember that imbalanced muscle strengthening can create serious problems. If we consider that our stretched muscles have elastic qualities, and that, in general, every muscle that produces action has an antagonist that produces a contrary action, we will understand that the predominance of one muscle over another can involve imbalances.

Let's take an example: muscles are like rubber bands. An individual who diligently trains his biceps, which bend the elbow, but ignores his triceps, which straighten the elbow, will have a thick, powerful rubber band pulling on the front of the forearm and a fragile, thin one pulling on the back. As a result, the elbow will display a significant degree of permanent flexion that will increase if the biceps is not stretched and the person does not perform work to produce muscle compensation. The same thing happens with individuals who develop their pectorals and ignore their back muscles: oftentimes, they assume a hunchback posture with a sunken chest.

This phenomenon also occurs naturally in our bodies, because there are groups of muscles that are much stronger than their antagonists, whether or not they are trained; one example is the quadriceps, in comparison to the ischiotibials. As a result, in application to athletics, in the presence of sudden contractions of the former, as in taking a shot on a soccer goal, the ischiotibials may be injured by the tension to which they are subjected. This can be avoided by stretching the latter muscle group to expand its range of motion before reaching its limit.

On the other hand, certain athletic pursuits, such as sprinting and weightlifting, tend to compress the joints, especially the intervertebral discs, and stretching after a workout can help remedy this situation.

Stretching may also help to reduce muscle tone. Yes, it can reduce muscle tone. This statement may appear crazy. Who wants to reduce muscle tone? Well, all of us should want to do this when a workout is over. Despite the fact that there is a persistent idea in many minds that muscle toning is important, it is produced only during exertion, and thereafter

Strengthening the biceps and allowing the triceps to remain weak is one form of undesirable muscular imbalance.

Taking a shot on goal involves sudden contraction of the quadriceps and subjects the ischiotibials to great tension.

the muscle should relax. Increasing muscle tone is an immediate response to physical activity, and, once this stops, the muscle should relax. If we want smoother, highly defined muscles, we can achieve this only by increasing their size and strength, because maintaining elevated muscle tone at rest is synonymous with strain or muscle contraction; it produces pain and certainly is not healthy.

So stretching exercises are the best way to achieve proper flexibility for performing a broad range of movements, for relaxing the muscles, and for improving our athletic performance. These exercises require minimal effort and do not produce fatigue, but rather relaxation and a sensation of lightness.

It goes without saying that achieving good flexibility requires time and dedication, as with everything worthwhile, and no miraculous gadget, whether it's a bracelet, a sticker with a hologram, or a highly advanced apparatus supported by hundreds of pseudoscientific studies, will improve our flexibility significantly in two weeks, much less instantaneously. Perseverance is the key to success.

The Concept and Methods of Stretching

The body reacts to every stress and stimulus that we apply to it. Thus, an individual who regularly lifts weights according to proper guidelines will experience muscle strengthening, a person who runs regularly will experience progressively greater stamina, and a person who regularly swims underwater will see that each time it is possible to stay underwater longer without breathing.

In the same way, a person who stretches regularly will achieve a response in the form of greater flexibility. This occurs because the body perceives the stimuli that are applied to it, and when these are sufficiently intense to produce stress, the body reinforces itself in order to confront a subsequent stimulus with greater guarantees and less upset to its balance. Similarly, when we come down with the flu our bodies strengthen their defenses, so we are not likely to experience it twice in the same winter. But why do we come down with the flu again the following winter? This can be explained by several factors, but one of them is that the body tends to economize

resources. If the flu does not attack again during the spring, summer, and fall, it does not appear necessary to maintain an elevated level of defense. Likewise, what's the point of maintaining big muscles, with the energy resources they consume, if we don't have to lift weights?

The same thing happens with sports in general, and with flexibility in particular. While we do stretching exercises and our bodies have to respond regularly to this muscle tension, we maintain optimal flexibility; but, if we take a few months off, whenever we return to training, we see that our flexibility has dropped off markedly.

Stretching beyond our limits means exposing ourselves to certain injury.

So the body needs a regular and adequate stimulus to improve and strengthen itself, and an excessive stimulus may exceed its capacity for recuperation and may lead to fatigue or injury. We have said that muscles have elastic properties—in other words, that we can change their shape by stretching or contracting them, and then they will return to their original shape. But what happens if we stretch excessively? It's very simple: if we exceed a muscle's elastic capacity and continue stretching it, the deformation will turn from elastic to plastic—that is, the muscle will continue to change shape, but it will no longer be capable of recovering its original shape and we will experience an elongation, a rupture of fibers, or a similar injury. This is why it is so dangerous to stretch to the point of pain. This applies to any athletic pursuit: when there is some resistance or ultraresistance athletes exceed their capacity for recovery or dealing with excess stimulus, they may suffer cardiac ailments, including heart attacks, and, in the worst case, death.

Therefore, it's good to be aware that our bodies cannot withstand everything and that it is important to know oneself and the sensations experienced during exercise. It may seem like a contradiction, but the most important thing in any physical activity is to use your head.

Stretching performed correctly will help us to improve many aspects of our fitness and will provide benefits in different situations:

Increase in Joint Mobility: Muscles may shorten for different reasons. Anyone who has experienced a fracture and had to keep an extremity immobile knows that, when the cast is removed, the joints cannot move normally, because they have lost a large portion of their range of movement. It is necessary to engage in rehabilitation sessions for several weeks to restore normal mobility. Muscle shortening can occur through inactivity, but also through significantly imbalanced activity. If we work on one muscle group regularly to strengthen it, but without performing broad movements, this muscle may shorten, and the problem is that it will be a very powerful muscle due to the strengthening, and stretching it may be more laborious. Thus, it is necessary to do regular stretching to keep the joints healthy and functional.

Blood Renewal: After performing an intense effort, the muscles tend to be congested, swollen, and tight. This is due to the accumulation of blood in the muscle during the exercise. To meet the greater requirement of energy and oxygen in the area being worked, the blood flow is increased to the muscles involved in the effort, because it's the blood that carries the glycogen and the oxygen to the tissues.

A body with very powerful pectorals, but weak back muscles, will tend to assume improper postures.

The problem appears when the draining of the blood is slower than the incoming supply, with a resulting accumulation in the muscles. The increased blood supply raises the blood pressure in the muscles, and this same pressure acts on the blood vessels, reducing the amount of blood that they can drain. We can see this more clearly with an example: if we try to suck drinks through straws while we squeeze the straws with our fingers, the drink will not reach our mouths, or it will flow in a much restricted quantity. Similarly, in the time immediately following a workout, it will be more difficult for the oxygen and the adequate nutrients to reach the muscles. At the same time, it will be difficult to remove the waste products generated during the exercise. Stretching exercises after the effort will help to drain that accumulated blood and allow new blood to reach the tissues, so recovery will be better and faster. In cases where athletic practice is divided into various periods, stretching during the breaks can also improve performance when the activity resumes.

Relaxing Strained Areas: We often experience strains in our backs, necks, and shoulders due to work, long trips in an automobile, carrying heavy items, bad posture, and so forth, and this translates into pain and discomfort. These can result from continuous, prolonged tension in a particular area of the body, which additionally has not been properly trained to withstand the tension. In this case, stretching will also soothe the areas that are strained, tense, and excessively tight. If the excessive effort cannot be avoided, it's a good idea to do some additional strength work with the affected muscles.

Balance Between Antagonistic Muscle Groups: Most posture problems are directly related to muscle imbalance. People with well-developed pectorals, but back muscles that are not similarly developed, tend to slouch and sink their chests. The same may occur with the lumbar region by strengthening and shortening the psoas major and the iliac, which is accompanied by weakness in the abdominal

Improper postures stress muscles and joints.

wall. In these cases, the best option is to stretch the muscle that has shortened or has become developed more than its antagonist, and to strengthen the latter in order to recover postural balance.

Preparation for Athletics: It is appropriate to include stretching in any good warm-up session, especially if it is done prior to competition or a demanding workout. As well as exercising the muscles, stretching brings beneficial effects in helping to warm them up, and it increases their viscosity and irrigation levels, and thus it improves athletic performance and minimizes the risk of injury. Stretching generally scheduled toward the end of the warm-up encourages blood renewal, and, especially, a slight increase in the range of joint motion through an increased stretching capacity of the muscles, which will allow the muscles to use this extra range to avoid an injury in case of a forced movement, a bad fall, a sudden change of direction, inadequate support, or maximum exertion. If we twist our ankles and our peroneus muscles are capable of stretching only 1.5 inches (4 cm), a twist that exceeds that range we will cause a sprain. On the other hand, if we use stretching exercises to allow our peroneus muscles to stretch about 2.4 inches (6 cm), we create an additional margin. This may not seem like much, and clearly stretching will never be a sure defense against injuries, but that extra range is significant in a twist and it may make the difference between a slight pain in the ankle for a few minutes and a serious injury that will keep one sidelined for weeks or months.

Once we know the purposes and the benefits of training for flexibility, it is also appropriate to explain that there is no single way to do stretches. In fact, there are several main methods, and every one of them can incorporate small variations.

Stretching helps avoid injuries in demanding sports and sudden exertions.

Static Stretches

In doing static stretches, we move slowly until we reach a position in which the muscles are stretched, and we hold it for a certain time. Normally, it is recommended to hold the stretching position for 15 to 60 seconds, and to do several repetitions of each exercise. Obviously, the time and the number of repetitions are defined by various factors, such as the person's objectives, performance level, the size and the strength of the muscle and its insertions, and the joints involved. This type of stretching is the most common, and it is the main focus of this work for various reasons. The first is that static stretches are done at a slow, controlled speed, and this allows us to work under extremely secure conditions. The slower a movement, the more difficult it is for us to lose control of it, and our ability to react in time to pain or discomfort is greatly increased. Along with the simplicity of the exercises and techniques used, this makes the stretches beneficial for both advanced athletes and people who are just beginning. In addition, the effectiveness of this type of stretching is amply corroborated, and the slowness of the movements and the static positions help limit the effect of the myotatic reflex. This reflex is a defense mechanism caused by the contraction of a muscle when it experiences a sudden stretching. In normal activities, this reflex protects the muscle from an injury as a result of stretching too far; but, during flexibility work, it can limit the performance obtained from the training and even cause a contrary effect. Static stretching surely has many benefits in its favor and very few strikes against it, and it is highly recommended for all types of practitioners, regardless of their level. In addition to basic static stretches there are other types.

Active Stretches: The stretching of a muscle results from the action of its antagonist. For example, when we stretch and move our arms rearward, the chest muscles stretch as a result of the contraction of the back muscles. This method is useful because, while one muscle stretches, its antagonist is working with contraction. Additionally, this is a fairly secure way to stretch, because the movement is done slowly, and it is difficult to stretch aggressively enough to cause injuries in a muscle simply through the progressive contraction of its antagonist. On the other hand, in this case, the myotatic reflex may come into play, and some muscle groups are not capable of stretching their antagonists enough to produce the optimal results, plus other mechanical factors can limit the stretch

Static stretching is simple in execution and requires no specific equipment.

and reduce the effectiveness of the exercise. For example, let's consider the ischiotibial muscles and their antagonist, the quadriceps: the latter is much more massive and powerful than the former, so it will be difficult to produce an effective stretch in the quadriceps by using only the contraction of the ischiotibial muscles.

Help from a companion in performing passive stretches contributes to greater and faster improvement.

Passive Stretches: The person doing the stretching invests no action or effort, but uses another person or something to produce the stretching position. These exercises are common in group workouts, in pairs, or with a trainer. They are very effective, because oftentimes the help from a companion or a stretching device makes it possible to go a little farther than we would working alone. Normally, it is possible to achieve slightly greater ranges of movement than those produced by active stretches, and thus a greater and quicker progression. Still, this method also has some drawbacks, such as the possibility of activating the myotatic reflex, and the greater risk of injury in comparison to static and active stretching. This greater risk is due to the fact that the individual is no longer in charge of the stretch, so there is a loss of control over the stretch and its consequences. If an assistant stretches us, no matter how careful this assistant may be (and it is necessary to have a careful partner), the sensitivity and the ability to react to our bodies that we have will not be there. There is always a risk of pulling a little too far or releasing the tension a little too late. In order to minimize these risks, the movement must be done very slowly, and there must be constant communication between the practitioner and the assistant.

Proprioceptive Neuromuscular Facilitation: This technique, which usually is done with an assistant, involves performing an isometric contraction of the muscle when it is stretched to the maximum. The assistant brings the muscle group to the stretching position. At that point, the stretched muscle is contracted isometrically for a few seconds—in other words, without producing any movement or shortening. Then the muscle is relaxed and the stretching position is increased slightly. This method, which originally was used in rehabilitation, has proven to be very effective; but, it probably involves a greater risk than the methods explained previously, and it's a good idea to do it with a qualified person.

Dynamic
Stretches

Dynamic stretches are done using controlled bouncing or swinging motions. For example, when upper body bends are done from side to side, we are working with dynamic stretching. Although in this type of stretching we seek to reach the end of the movement and perform a little bounce, the speed of the movement must be moderate and the bounce must be controlled. If these basic guidelines are observed, the possibility of experiencing an injury will be greatly reduced and we can work safely, although not at the level of static stretches. Some studies indicate that dynamic stretches contribute to athletic performance to a greater degree, specifically because the movement factor is present in every physical activity. Nevertheless, if our goal is to reach a heightened level of flexibility and significantly increase our range of movement, static stretching and proprioceptive neuromuscular facilitation will contribute to greater and faster improvement. There is also a variant in dynamic stretching:

Dynamic stretching is a good choice during a warm-up before an athletic practice.

Ballistic Stretches: As with dynamic stretches, the goal is to improve flexibility through movement; but, in this case, the movement is done at greater speed and a bounce is used at the end of the motion. In general, momentum is applied to an extremity to move it at high speed to its limit, at which point there is a bounce. Although it is still common to see this type of stretching in workout sessions with all kinds of athletes, little by little, and particularly with the emergence of physical coaches and trainers as professionals, it is used less frequently than in the past. This falling out of use is due to various factors: the most significant one is the high risk of injury involved. High-speed stretches involve less control over movements and the ability to stop them before an undesirable sensation: the damage is already done by the time you stop the inertia of the movement. In addition, the myotatic reflex produced in dynamic stretches is more pronounced in ballistic stretching, and this can have adverse effects. Finally, the speed at which the stretching is produced means that its effect on flexibility is minimal. Thus, this method provides meager improvement with a high risk of injury.

The sudden movement of ballistic stretching makes it a poor choice for the general public, because of the high risk of injury involved.

Basic Stretching Principles

We must keep many factors in mind before beginning to stretch. Although we have already discussed some of them, it is appropriate to review them thoroughly.

Our bodies can bring us many joys, but also many problems. In general, we don't appreciate the possibilities our bodies present: we take them for granted. We start to realize what we have at the point where we run the risk of losing it, or when we realize we have already lost it. We don't realize how much smoking harms us until the elevator breaks down and it takes a huge effort to climb the stairs. In the same way, we are unaware of the poor state of our muscles until one day we get up with a painful muscle cramp that reminds us for several weeks that we should have started taking care of ourselves long ago. Probably before reaching these straits our bodies have sent us signals about what was in store for us; but, because the problems were minor or responded to anti-inflammatory medications, we didn't pay much attention. Our bodies commonly talk to us, telling us through symptoms and sensations about what is happening to them. That's the way it works in daily life, including during exercise. In fact, many Asian and relaxation disciplines emphasize getting to know yourself. If we apply this to flexibility work, we improve more quickly and we avoid injury. In the first place, we need to banish the myth of no pain, no gain, and to apply the principle that progress comes through effort and perseverance.

During flexibility training, we need to be conscious that a sense of discomfort and tension in the muscle being stretched must be present. The body perceives this stimulus, and, by means of this sensation, it indicates that the stretch is under way. On the other hand, if we feel pain, we need to interpret the message differently: pain almost always means injury or risk of injury, and we need to reduce the intensity of the stretch.

Breath control is basic to all athletic disciplines, but it takes on special importance in flexibility work, because it contributes to relaxation, one of the purposes of stretching. Rapid, superficial breathing and interrupted breathing will scarcely bring us to a state of calm; but, deep, deliberate breathing can do so. This must be the general principle that we apply to breathing during flexibility work, although there are isolated exceptions, because the performance of certain exercises requires the expulsion of air from our lungs at a certain time or superficial breathing to achieve optimal results.

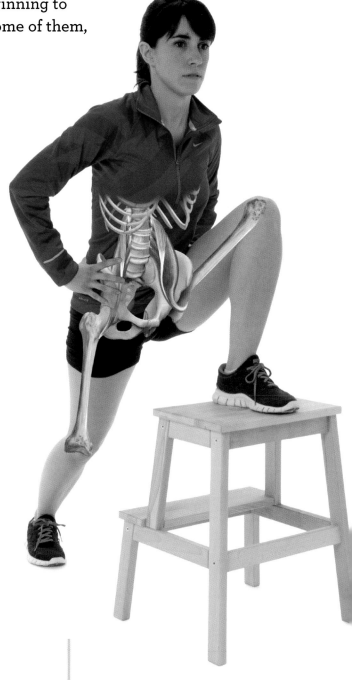

A sensation of discomfort, but not pain, is necessary while stretching.

We also need to remember that stretching exercises need to be done slowly and progressively in order to provide a greater margin in perceiving sensations and acting accordingly. If we perform an exercise at high speed, as with ballistic stretches, there is no margin of reaction between the first pain signal and an injury, because they will occur at practically the same instant. On the other hand, after the first sensation of pain in a stretching exercise performed slowly, we will have adequate leeway in which to stop the exercise before the muscle or some joint is injured. We need to pay attention to these sensations in the joints that are unstable, such as the shoulder, because they can dislocate relatively easily, especially in people who have previously had problems of this type.

Also, we need to be conscious of the fact that our bodies work better when warmed up: they perform better and are less susceptible to breakage. Just as a motor cannot perform at peak until it is warmed up, we cannot stretch our muscles before warming up. When we are cold, we are stiffer and more fragile, which is why stretches should be done toward the end of the warm-up, at the end of the athletic practice, or at both times. Certainly, we have sometimes seen people who go out for a run and stretch before they begin. Now, we know that this is not the best way, and it would be better to stretch after several minutes of gentle jogging and at the end of the run.

Finally, we need to very careful with the position of the spine and always be aware of it. Oftentimes, in our zeal to go farther in a stretch, we subject our spines to excessive tension. We must avoid this tension and always be attentive to our position so that our backs are straight whenever the exercise allows. It is clear that certain stretches, especially those that affect the muscles of the upper body and the neck, produce unavoidable changes in the position of the spine. This is not bad, but we need to be careful.

We want to point out that, when we say that the spine must be straight, we are referring to the fact that it must be curved, but without exceeding its natural curvature. Basically, there are three totally normal curvatures in the spine: one is the curvature toward the rear in the dorsal area, known as kyphosis; the two others are curvatures toward the front that are found in the cervical and lumbar areas, which are referred to as lordosis. In no case, should we attempt to eliminate or reduce them. The three curvatures are necessary for good postural health, and they need treatment only if they are overly exaggerated, which can give rise to as many problems as a lack of them can.

A healthy spine must preserve its natural curvature, but without exaggerating it.

✔
NORMAL SPINE

✗
STRAIGHTENED SPINE

✗
HYPERKYPHOSIS

✗
CERVICAL HYPERLORDOSIS

✗
LUMBAR HYPERLORDOSIS

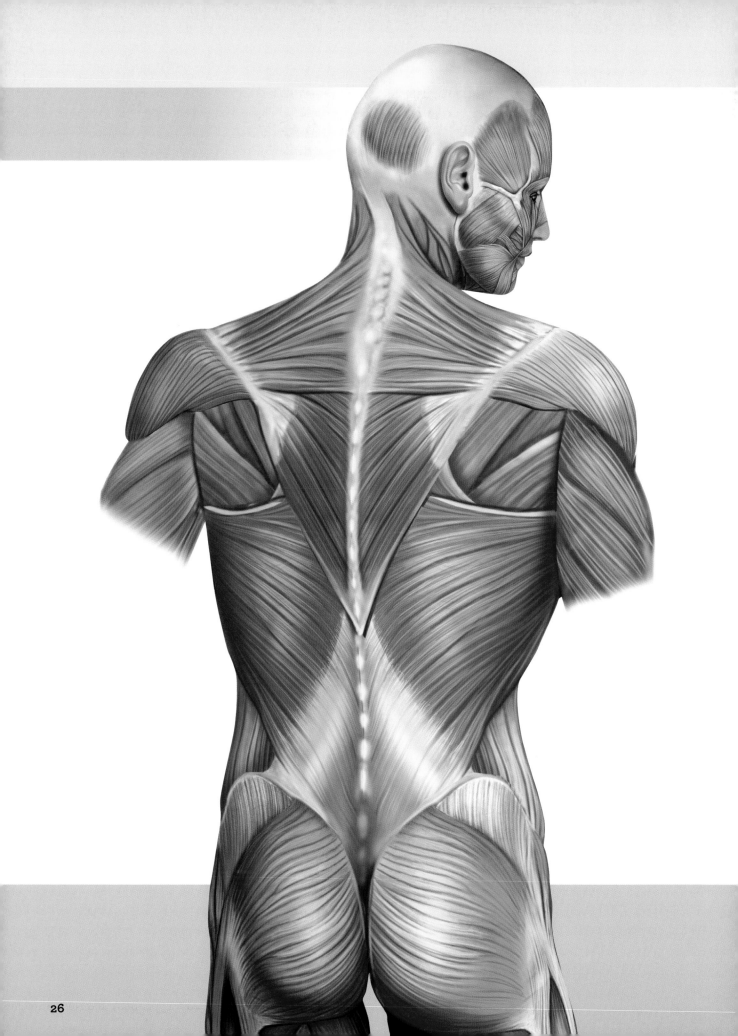

UPPER BODY AND
NECK STRETCHES

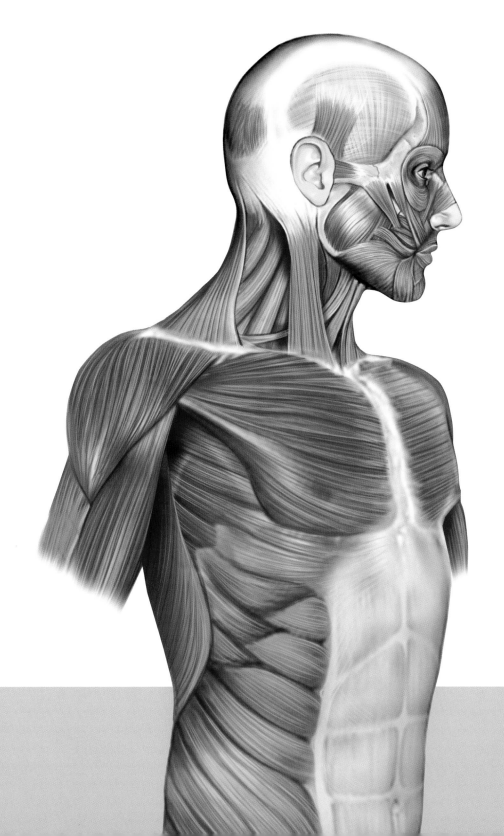

NECK STRETCHES

TRAPEZIUS

This muscle originates at the occipital bone, on the spiny process of the cervical and thoracic vertebrae, and it inserts at the acromion and the spine of the scapula. The origin is fan-shaped, so its various portions perform different functions, depending on the direction of the fibers: the upper portion elevates the scapula; the middle produces adduction; and the lower part, depression.

LEVATOR SCAPULAE

This muscle originates at the transverse processes of vertebrae C1 through C4, and inserts at the upper medial edge of the scapula. As the name indicates, its function is to raise the scapula.

SCALENES

There are three scalene muscles: the anterior, the middle, and the posterior.

Scalenus anterior (anterior scalene): It originates at the transverse processes of vertebrae C3 through C6, and it inserts at the inner edge of the first rib. Its function is flexion and lateral flexion of the neck, as well as elevation of the first rib.

Scalenus medius (middle scalene): This muscle originates at the transverse processes of vertebrae C2 through C7, and it inserts at the upper surface of the first rib. Its functions are the same as those of the anterior scalene.

Scalenus posterior (posterior scalene): This muscle originates at the transverse process of vertebrae C4 through C6, and it inserts at the outer surface of the second rib. Its function is flexion and lateral flexion of the neck, as well as elevation of the second rib.

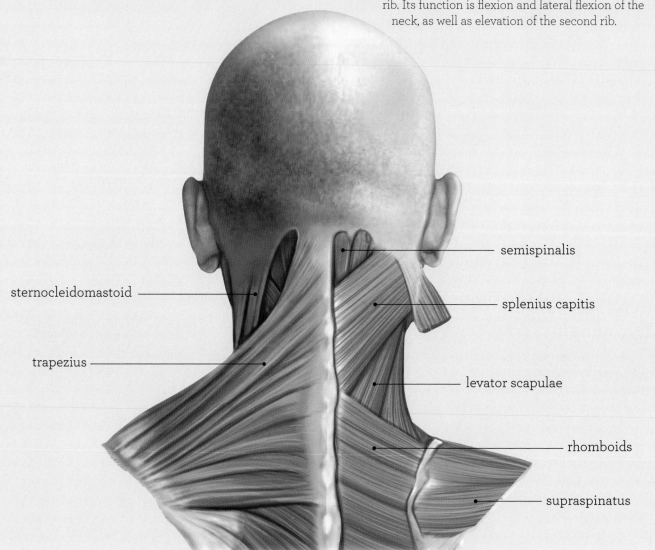

sternocleidomastoid

trapezius

semispinalis

splenius capitis

levator scapulae

rhomboids

supraspinatus

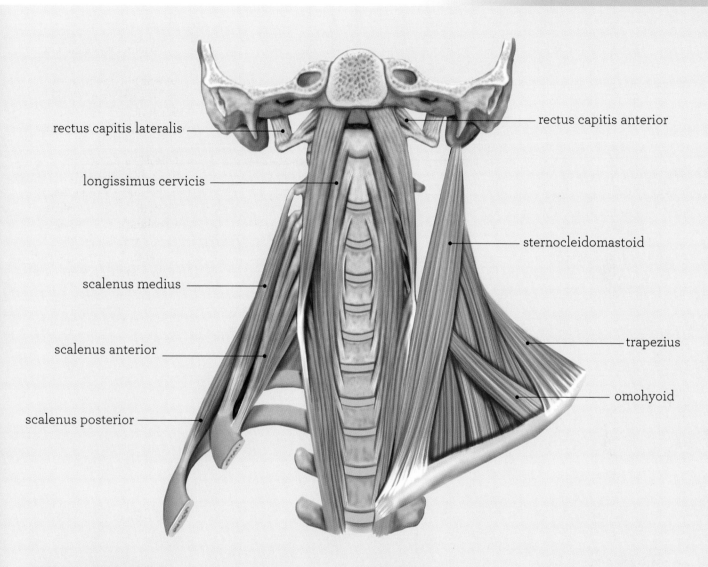

rectus capitis lateralis

longissimus cervicis

scalenus medius

scalenus anterior

scalenus posterior

rectus capitis anterior

sternocleidomastoid

trapezius

omohyoid

STERNOCLEIDOMASTOID
The sternal portion of this muscle originates at the sternum, and the clavicular portion at the medial third of the clavicle. Both portions have a common insertion at the mastoid process of the temporal bone. This muscle's functions are neck flexion, lateral flexion, and rotation.

SPLENIUS
There are two splenius muscles: the splenius capitis and the splenius cervicis. However, we will treat them as one in the stretches, because they have functions and methods of stretching in common.
Splenius capitis: This muscle originates at the nuchal ligament, from vertebra C3 to C7, and the spiny processes of vertebrae C7 to T4. Its insertion is located at the mastoid process of the temporal bone and at the occipital bone. Its functions are neck extension, lateral flexion, and rotation.
Splenius cervicis: This muscle originates at the spiny processes of vertebrae T3 through T6, and it inserts at the transverse processes of vertebrae C1 through C3. Its functions are neck extension, lateral flexion, and rotation.

Lateral Neck Bend

START

Stand with your arms at your sides and your feet shoulder width apart, and relax the muscles of your scapular belt.

TECHNIQUE

While looking to the front, tilt your head to one side by means of lateral neck flexion, as if you were trying to touch your ear to the shoulder opposite the stretch. Hold the position without relaxing, and feel the stretch in the trapezius and the muscular discomfort that it produces. If you want a little more intensity, lower the shoulder on the side being stretched.

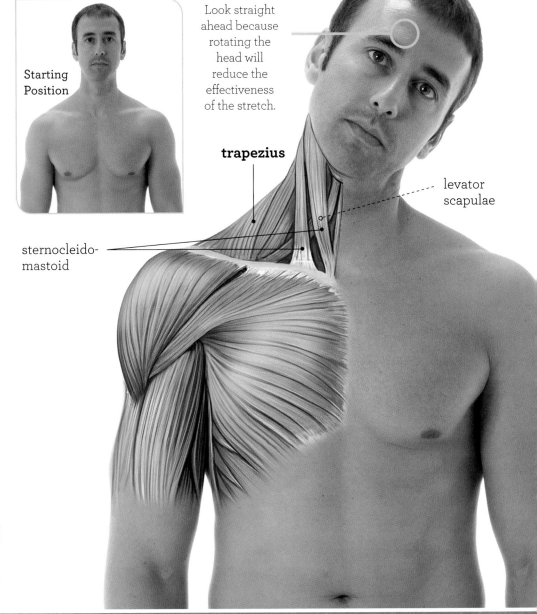

Starting Position

Look straight ahead because rotating the head will reduce the effectiveness of the stretch.

trapezius

levator scapulae

sternocleido-mastoid

LEVEL	REPS	DURATION
BEGINNER	3	20 sec.
INTERMEDIATE	2	30 sec.
ADVANCED	4	35 sec.

CAUTION	BENEFITS	INDICATION
Remember that the sensation during the stretch should be discomfort, but not pain, especially when you are moving your cervical column.	Reduction of tension in the posterior area of the neck, an area that is sensitive to tensions created by physical exertion, stress, and certain sustained postures.	For anyone who experiences discomfort and pain as a result of the position assumed during work, especially in administrative and office work.

Assisted Neck Bend

Keep looking ahead.

levator scapulae

sternocleidomastoid

trapezius

Starting Position

START
Take a position with your back straight and one hand on your head. Look toward the front without flexing or extending your head.

TECHNIQUE
Perform traction on your head, with your hand contacting it as if you were trying to move your ear to your shoulder, in order to produce lateral flexion of the neck and head. Hold the position during the time appropriate to your level. You can place your free arm behind your back or lower your shoulder on the side being stretched in order to increase the intensity.

LEVEL	REPS	DURATION
BEGINNER	2	20 sec.
INTERMEDIATE	3	30 sec.
ADVANCED	4	35 sec.

CAUTION
In order to protect your cervical vertebrae, avoid applying excessive traction with your hand. Remember that you must not feel joint pain, but only a muscle discomfort resulting from the stretch.

BENEFITS
Reduction of tension in the posterior and lateral regions of the neck.

INDICATION
For people who experience pain in the posterior and lateral regions of the neck, mainly as a result of their jobs, especially when those jobs require sitting in front of a computer, a desk, or a table for many hours.

Neck Bend and Rotation

START
Sit or stand with your spine straight and your gaze directed toward the front. Let your arms hang at your sides and relax your shoulders.

TECHNIQUE
Turn your neck at least 45° and lower your chin as if you were trying to touch it to your chest. This movement will produce muscle tension for the stretch, and you will note a slight upward movement of the shoulder blade opposite the direction of your chin. Hold this position for the time appropriate to your level, and then do the exercise toward the other side.

During the exercise, remember to relax your shoulders so they remain loose and there is no other tension in the muscles in the area other than that created by the stretch.

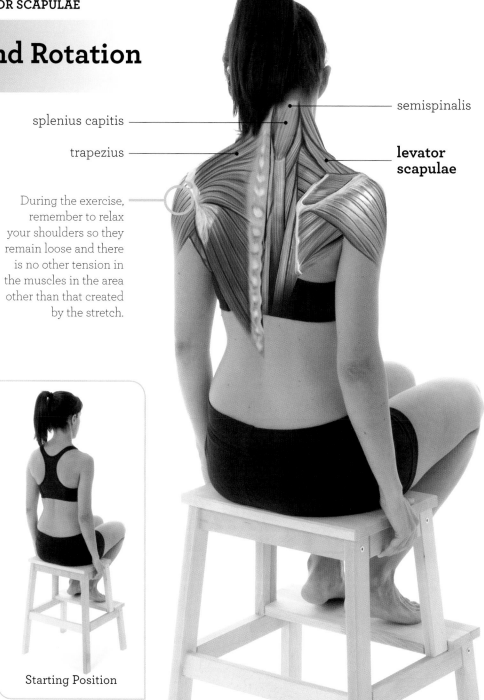

semispinalis

splenius capitis

trapezius

levator scapulae

Starting Position

LEVEL	REPS	DURATION
BEGINNER	2	20 sec.
INTERMEDIATE	3	25 sec.
ADVANCED	4	30 sec.

CAUTION
Remember to keep your back relaxed, but straight and perpendicular to the floor.

BENEFITS
Reduction in muscle tension in the whole posterior region, and thus the pain that excess tension in this area can produce.

INDICATION
For people who spend many hours seated in the workplace, especially in administrative tasks.

Neck Rotation and Extension

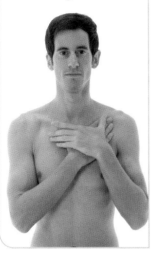

Starting Position

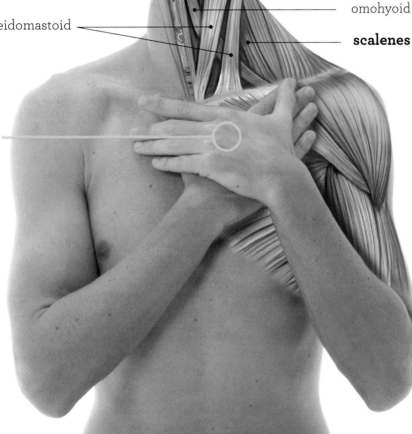

sternocleidomastoid

omohyoid

scalenes

Use hand pressure to keep your chest from rising and to make the stretch more effective.

START
Stand or sit with your back straight and your hands on the upper part of your chest, near the base of your neck and slightly to the side of the muscle to be stretched. The hands should be placed one on the other and should exert pressure inward and downward.

TECHNIQUE
Extend your neck and point your face away from the side where your hands are placed, while keeping pressure on your chest with your hands. At the point of maximum stretch, you will note that the chest area where your hands are placed will tend to raise slightly because of the tension that the scalenes produce on the first and second ribs. Hold this position of maximum stretch for a few seconds and repeat the exercise on the other side.

CAUTION	**BENEFITS**	**INDICATION**
Exercises that produce neck extension may cause discomfort in the cervical area in the presence of preexisting problems. Proceed slowly and gradually so you can stop in time if you feel discomfort.	Reduction of tension in the whole anterior neck region. These muscles do not cause as much discomfort as the posterior ones, but it's a good idea to stretch them preventively.	For people who have been diagnosed with compression of the brachial plexus, for reducing tension in the muscles that cause it, plus for people who spend many hours seated in front of a computer.

LEVEL	REPS	DURATION
BEGINNER	2	15 sec.
INTERMEDIATE	3	20 sec.
ADVANCED	4	30 sec.

Neck Extension and Chin Raise

START

Stand or sit with your spine straight and your gaze directed straight ahead. Place your hands on the upper part of your chest slightly to one side, as in the previous exercise. Apply pressure to your clavicle on the side of the muscle being stretched.

TECHNIQUE

Extend your neck and direct your gaze upward. Then move your head to the side as if you wanted to look at the ceiling with the eye on the side being stretched. At the same time, maintain pressure with your hands to keep your clavicle from rising with the tension produced in the stretch. Hold the stretch for the time appropriate to your level, and then repeat the stretch on the other side.

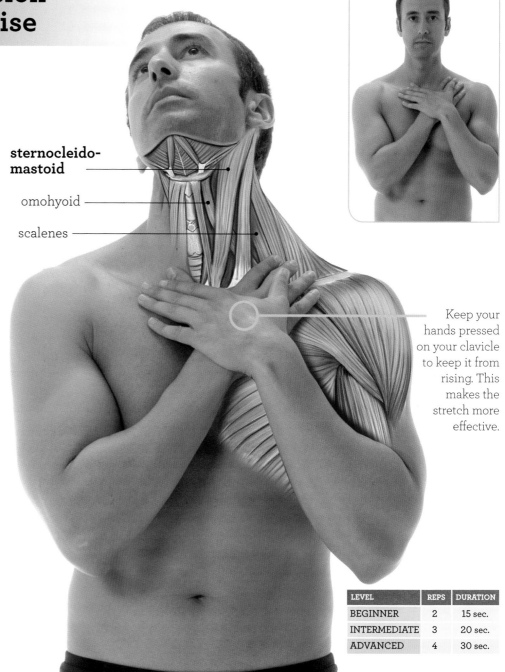

Starting Position

sternocleido-mastoid

omohyoid

scalenes

Keep your hands pressed on your clavicle to keep it from rising. This makes the stretch more effective.

LEVEL	REPS	DURATION
BEGINNER	2	15 sec.
INTERMEDIATE	3	20 sec.
ADVANCED	4	30 sec.

CAUTION

Keep your shoulders relaxed and avoid raising them during the stretch. Do not do this exercise if you feel discomfort in the cervical area.

BENEFITS

Relaxation of the anterior neck region.

INDICATION

For people who experience compression of the brachial plexus, and who spend lots of time seated or lying down, such as administrative workers, professional drivers, or convalescents. The latter should consult their doctors.

Assisted Neck Bend

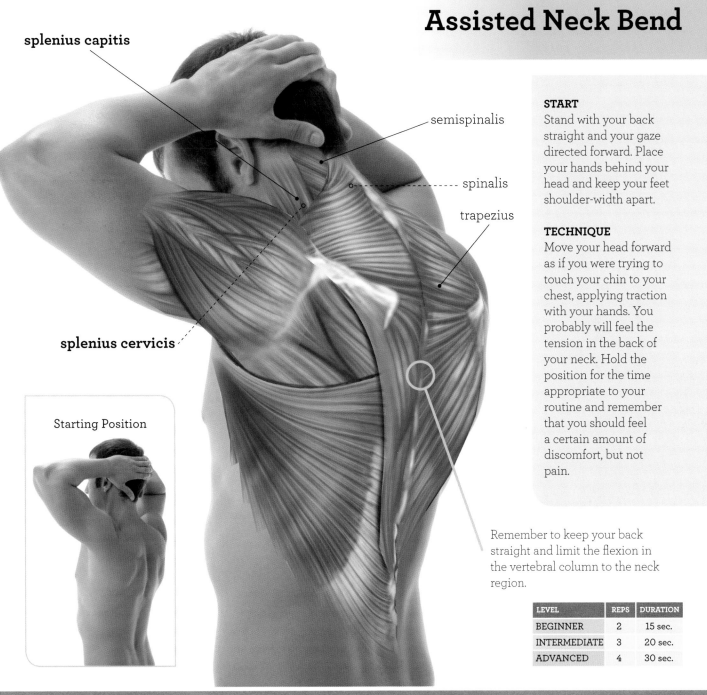

splenius capitis

semispinalis

spinalis

trapezius

splenius cervicis

Starting Position

START
Stand with your back straight and your gaze directed forward. Place your hands behind your head and keep your feet shoulder-width apart.

TECHNIQUE
Move your head forward as if you were trying to touch your chin to your chest, applying traction with your hands. You probably will feel the tension in the back of your neck. Hold the position for the time appropriate to your routine and remember that you should feel a certain amount of discomfort, but not pain.

Remember to keep your back straight and limit the flexion in the vertebral column to the neck region.

LEVEL	REPS	DURATION
BEGINNER	2	15 sec.
INTERMEDIATE	3	20 sec.
ADVANCED	4	30 sec.

CAUTION

The cervical column is delicate, so you need to exercise particular care to avoid applying excessive traction with your hands.

BENEFITS

Reduction of tension in the posterior region of the neck.

INDICATION

For people who experience discomfort in the back of the neck, especially those who spend many hours at work seated in front of a computer, at a table, or behind a steering wheel.

DORSAL STRETCHES

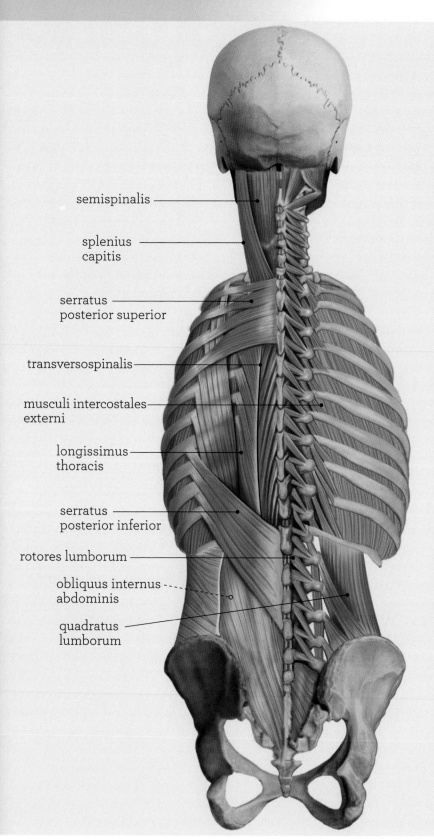

semispinalis

splenius
capitis

serratus
posterior superior

transversospinalis

musculi intercostales
externi

longissimus
thoracis

serratus
posterior inferior

rotores lumborum

obliquus internus
abdominis

quadratus
lumborum

SERRATUS ANTERIOR
This muscle originates in the upper nine ribs, and it inserts at the medial edge of the scapula. Its functions are raising and lowering the scapula, plus attaching it to the upper body, and it plays a part in inhalation.

LATISSIMUS DORSI
This muscle originates at the spiny processes of vertebrae T6 through L5 and the vertebrae of the sacrum and at the posterior ridge of the ilium, and it inserts at the proximal third of the humerus. Its main function is extension and retropulsion of the shoulders, so it is one of the climbing muscles, although it also produces adduction and medial rotation of the shoulder.

SEMISPINALIS
This muscle originates at the transverse processes of vertebrae T7 through T10 and at the articular processes of vertebrae C4 through C6, and it inserts at the occipital bone and the spiny processes of vertebrae C2 through T4. Its main function is the extension of the spinal column, and it also takes part in its lateral flexion.

RHOMBOIDS
There are two rhomboid muscles, the major and the minor.
Rhomboideus major: This muscle originates at the spiny processes of vertebrae C7 through T1 and at the nuchal ligament, and it inserts at the upper part of the medial edge of the scapula. Its function is to retract and raise the scapula.
Rhomboideus minor: This muscle originates at the spiny processes of vertebrae T2 through T5, and it inserts at the medial edge of the scapula. It shares functions with the rhomboideus major.

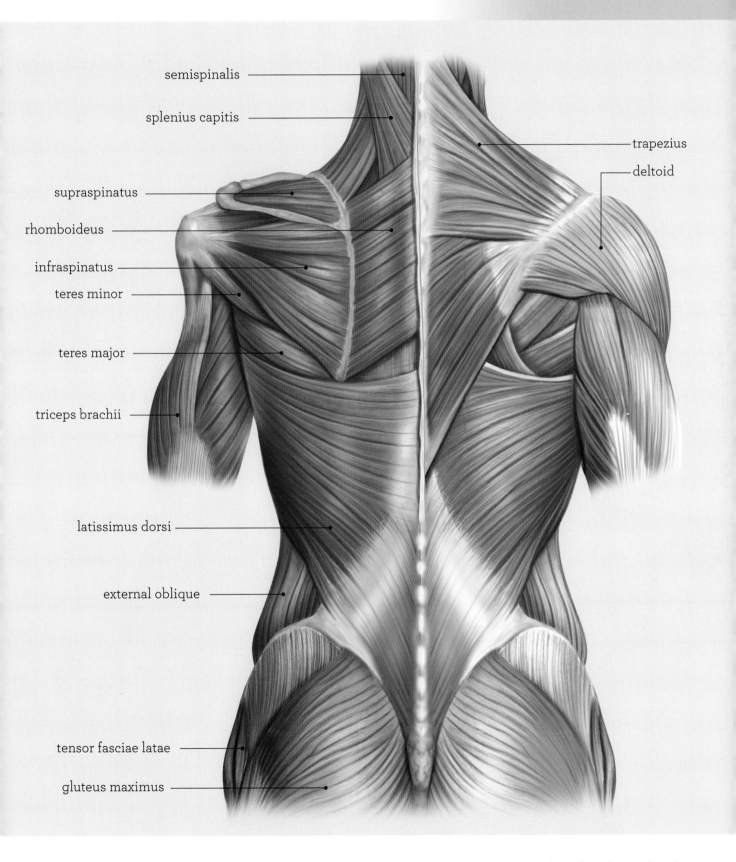

semispinalis

splenius capitis

trapezius

deltoid

supraspinatus

rhomboideus

infraspinatus

teres minor

teres major

triceps brachii

latissimus dorsi

external oblique

tensor fasciae latae

gluteus maximus

Hands over Your Head

START

Stand with your arms extended forward. Cross your hands and place your palms together as you interlock your fingers in such a way that your thumbs are pointing downward. Your knees need to be slightly bent, and your elbows totally extended.

TECHNIQUE

Raise your arms so that your hands are above your head. At this point, continue moving them rearward, until the muscle tension becomes obvious and you cannot go any farther. As you raise your arms, you may bend your elbows slightly so they don't touch your head.

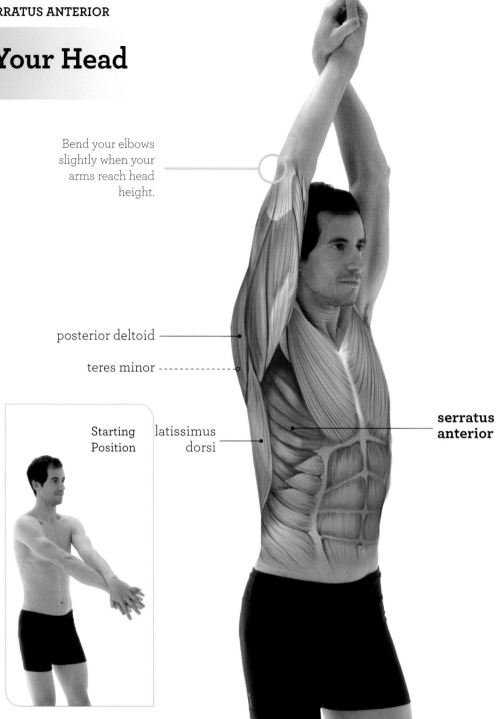

Bend your elbows slightly when your arms reach head height.

posterior deltoid

teres minor

Starting Position

latissimus dorsi

serratus anterior

LEVEL	REPS	DURATION
BEGINNER	3	20 sec.
INTERMEDIATE	3	30 sec.
ADVANCED	4	40 sec.

CAUTION

Avoid bending your torso to the rear to increase the range of your hands, because you would end up in an unstable position without adding to the stretch.

BENEFITS

Reduction of tension in the dorsal area of the back. Relief from tension related to carrying and holding heavy objects with the upper extremities, or from remaining seated for long periods of time.

INDICATION

For people who do office work and spend many hours seated at a desk or a computer, as well as for those who regularly handle fairly heavy items, especially if they need to bend over a work area, as care providers for very young children do.

Supine Shoulder Antepulsion

Starting
Position

LEVEL	REPS	DURATION
BEGINNER	3	20 sec.
INTERMEDIATE	4	30 sec.
ADVANCED	5	35 sec.

START
Lie down on your back on a mat, with your legs straight and together and your arms alongside your body. Keep your back straight and aligned with your legs.

TECHNIQUE
Using shoulder antepulsion, raise your arms above your head, as if you were trying to create the greatest possible distance between the tips of your fingers and your toes. Your arms should be parallel and as close to the floor as possible. The backs of your hands should contact the floor with your hands palms up.

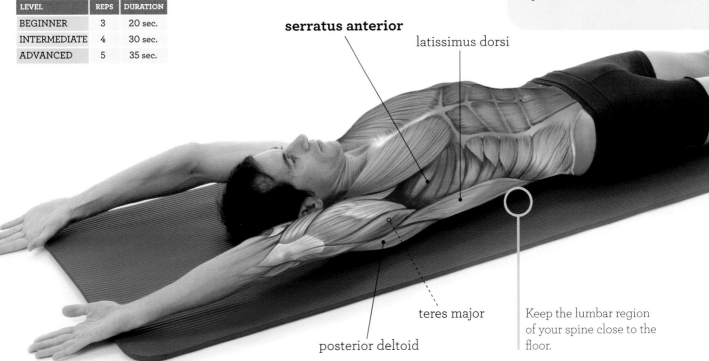

serratus anterior

latissimus dorsi

teres major

posterior deltoid

Keep the lumbar region of your spine close to the floor.

CAUTION

In this exercise, you need to be careful with your back. The natural instinct in lowering your arms to the floor is to accentuate the lumbar lordosis so that the spine is greatly arched. Avoid this and keep your lumbar area close to the mat. As with most stretches involving the shoulder joint, be careful if you feel discomfort, especially if you experience shoulder problems or have previously had a dislocation.

BENEFITS

Reduction of tension in the dorsal area. This area is sensitive to tension related to physical exertion, to carrying heavy weight, or to sitting for long periods of time.

INDICATION

For people who experience pain in the dorsal area as a result of their work posture, especially those who spend many hours in front of a computer or sitting. Also for people who do repetitive work with their arms and upper bodies, or who move or lift heavy objects.

Mohammed Position

START

Kneel down on a mat and lower yourself until you are sitting on your haunches. Lean your upper body forward and rest your hands on the mat, as shown in the starting position.

TECHNIQUE

Slide your hands forward over the mat, as if you were trying to reach the end with your fingertips. Lean your upper body forward and keep your elbows straight. When you lower your chest, your head will be between your arms and you will feel the tension from the stretch in your back and ribs.

Starting Position

LEVEL	REPS	DURATION
BEGINNER	3	20 sec.
INTERMEDIATE	4	30 sec.
ADVANCED	5	40 sec.

Keep your arms straight and parallel, and keep your heels in contact with your gluteals.

deltoid posterior

teres major

latissimus dorsi

serratus anterior

CAUTION

Once again, be careful with the shoulder joint, and use a padded mat to reduce pressure on your ankles at the beginning.

BENEFITS

Reduction of pain caused by excessive tension in the dorsal and lumbar areas.

INDICATION

For people whose work requires them to hold one position for many hours, whether standing or sitting. In jobs that require sitting at a desk or standing behind a counter, it will be helpful to do this stretch during breaks or at the end of the work day.

Crossed Arms

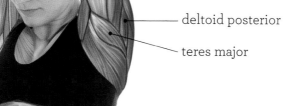

START
Stand with your arms raised and your forearms crossed above your head. Keep your back straight and look toward the front.

TECHNIQUE
Straighten your elbows and move your hands upward while crossing them, so that they remain linked without separating, no matter how much force you apply in straightening your elbows. Try to move your arms back slightly with respect to your body in order to make the stretch more effective.

deltoid posterior

teres major

serratus anterior

latissimus dorsi

Maintain the natural lumbar curvature without exaggerating it.

Starting Position

LEVEL	REPS	DURATION
BEGINNER	3	20 sec.
INTERMEDIATE	4	30 sec.
ADVANCED	5	40 sec.

CAUTION
Avoid arching your back as you try to move your arms rearward, because this would be detrimental to your back and would not improve the stretch in any way.

BENEFITS
Reduction of back tension.

INDICATION
For people whose work requires them to spend many hours sitting.

Raised Arm Upper Body Bend

START

Take a standing position and raise one arm through shoulder abduction. Bend your elbow and keep your palm facing forward as if you were giving the signal to stop. The opposite arm can remain relaxed at your side.

TECHNIQUE

Starting from the position described, continue raising your hand and move it slightly toward the front and toward the opposite side, while you bend your upper body to the side toward which your hand is moving. This gesture should be similar to a crawl stroke, but you hold it for a few seconds at the point of maximum extension in order to achieve a better stretch.

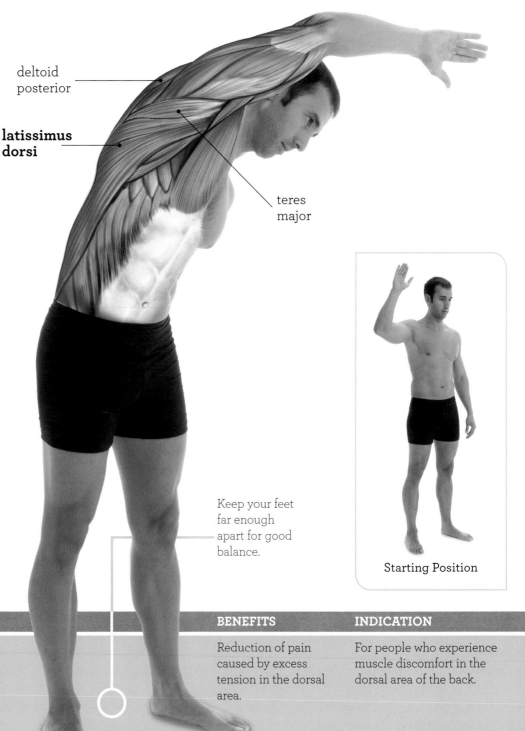

deltoid posterior

latissimus dorsi

teres major

Keep your feet far enough apart for good balance.

Starting Position

LEVEL	REPS	DURATION
BEGINNER	3	20 sec.
INTERMEDIATE	4	30 sec.
ADVANCED	5	40 sec.

CAUTION

This stretch involves no difficulty or risk of any kind as long as you start from a stable, balanced position, so keep your feet in line for a good support base.

BENEFITS

Reduction of pain caused by excess tension in the dorsal area.

INDICATION

For people who experience muscle discomfort in the dorsal area of the back.

Pulling from a Fixed Point

Starting Position

teres major

latissimus dorsi

deltoid posterior

serratus anterior

Keep your elbows straight during the entire exercise and bend your knees slightly if you feel much tension in the back of your legs or your lumbar area.

LEVEL	REPS	DURATION
BEGINNER	3	20 sec.
INTERMEDIATE	4	30 sec.
ADVANCED	5	40 sec.

START

Take a position in front of a support point that is no lower than your waist. You can use a tall stool, a table, a countertop, the back of a chair, etc. Stand far enough away from the support point so you have to bend your upper body forward and extend your arms to reach it. Hold the selected fixed point with both hands.

TECHNIQUE

Starting from the described position, try to lower your chest, keeping your arms extended. Go as low as you can without feeling pain, just tension, in your ribs, and hold the position for the time appropriate to your level.

CAUTION

As in many exercises involving a forced antepulsion of the shoulder, you must be particularly attentive to the sensations in your shoulders and immediately reduce the intensity of the stretch if you feel discomfort in those joints. It will also help to add a slight bend in the knees if you experience lumbar discomfort.

BENEFITS

Reduction of discomfort caused by excessive tension in the dorsal area.

INDICATION

For anyone who experiences pain or discomfort in the dorsal area due to strain, especially people who spend lots of time standing behind a counter.

Assisted Upper Body Bend

START
Sit on a stool or anything else with no back rest. Fold your hands and place them on the back of your neck, keeping your back straight.

TECHNIQUE
Bend your head, neck, and upper body so that you curve forward. You can apply gentle pressure to your head to maximize the effect of the stretch. Hold the point of maximum stretch for the appropriate time.

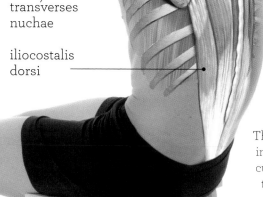

rectus capitis posterior major

rectus capitis posterior minor

semispinalis

transverses nuchae

iliocostalis dorsi

This exercise involves the curvature of the dorsal and cervical vertebrae.

Starting Position

LEVEL·	REPS	DURATION
BEGINNER	1	15 sec.
INTERMEDIATE	2	20 sec.
ADVANCED	3	25 sec.

CAUTION
Do not apply much pressure to your head, because the cervical section of the spine is especially sensitive to tension, and this could be counterproductive or even detrimental.

BENEFITS
Reduction of discomfort in the dorsal and lumbar areas when a significant portion of the muscles in this zone are stretched.

INDICATION
For people who experience discomfort in the dorsal and lumbar areas, mainly those whose work requires many hours of standing or sitting.

Arms Forward

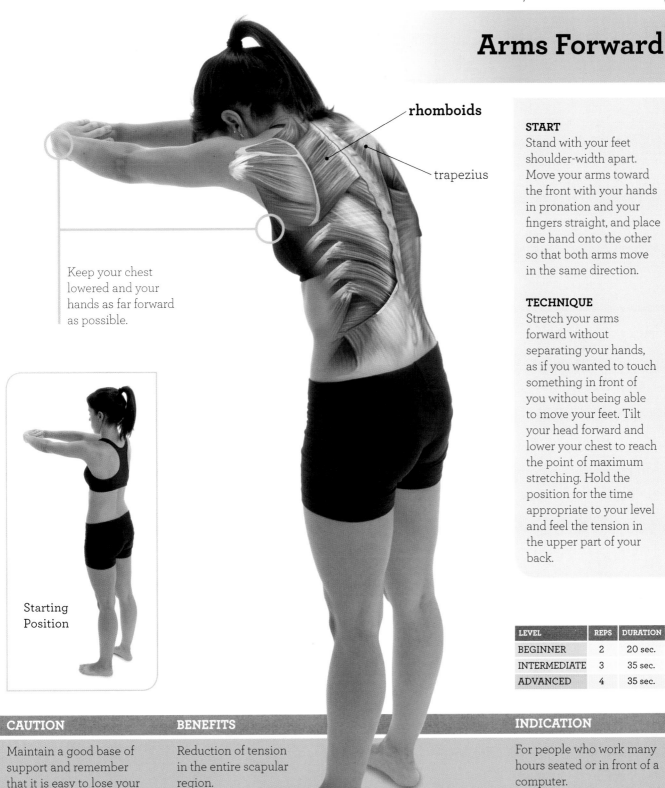

rhomboids

trapezius

Keep your chest lowered and your hands as far forward as possible.

Starting Position

START
Stand with your feet shoulder-width apart. Move your arms toward the front with your hands in pronation and your fingers straight, and place one hand onto the other so that both arms move in the same direction.

TECHNIQUE
Stretch your arms forward without separating your hands, as if you wanted to touch something in front of you without being able to move your feet. Tilt your head forward and lower your chest to reach the point of maximum stretching. Hold the position for the time appropriate to your level and feel the tension in the upper part of your back.

LEVEL	REPS	DURATION
BEGINNER	2	20 sec.
INTERMEDIATE	3	35 sec.
ADVANCED	4	35 sec.

CAUTION

Maintain a good base of support and remember that it is easy to lose your balance.

BENEFITS

Reduction of tension in the entire scapular region.

INDICATION

For people who work many hours seated or in front of a computer.

Upper Body Hug

START

Stand with your arms folded in front of your chest in such a way that each hand is on the rear of the opposite shoulder.

TECHNIQUE

Starting with the position described, try to reach the center of your shoulder blades with the tips of your fingers. Remember that each hand tries to reach the opposite shoulder blade over the side, not over the top, as if you were giving yourself a hug. When you reach the point of maximum tension, hold the position for the time appropriate to your level.

rhomboids

trapezius

Stretch your fingers around, not over, your shoulders.

Starting Position

LEVEL	REPS	DURATION
BEGINNER	2	20 sec.
INTERMEDIATE	4	25 sec.
ADVANCED	5	35 sec.

CAUTION

Remember to exhale progressively during the stretch, and breathe lightly and shallowly as you hold it. Lowering your chest will help you maximize the stretch, and holding your breath is counterproductive while performing any exercise.

BENEFITS

Reduction of tension in the entire scapular region.

INDICATION

For people who work sitting or in front of a computer for hours.

Leg Hug

Starting Position

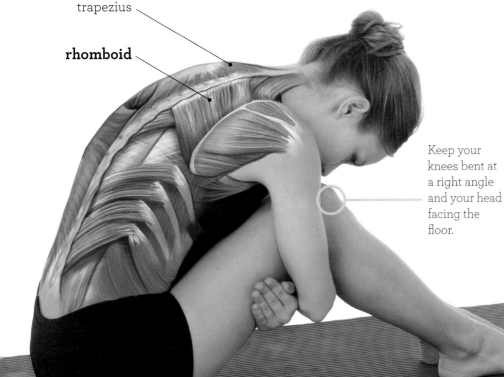

trapezius

rhomboid

Keep your knees bent at a right angle and your head facing the floor.

LEVEL	REPS	DURATION
BEGINNER	2	20 sec.
INTERMEDIATE	4	25 sec.
ADVANCED	5	35 sec.

START

Sit on a mat with your legs together, with the soles of your feet planted firmly and your knees bent at about 90°. Bend your upper body forward and hug your legs with your arms and hands.

TECHNIQUE

Contact your thighs with your chest and try to stretch the leg hug so that each hand reaches as high as possible on the opposite arm. As you do so, you will notice that your chest curves in and increases the tension of the stretch in the area of your shoulder blades and the spinal column.

CAUTION	BENEFITS	INDICATION
Do not hold your breath. Exhale progressively while reaching the point of maximum stretch, and then breathe shallowly until the end.	Reduction of tension in the scapular region.	For people who spend many hours seated in front of a computer, behind a counter, or at the steering wheel.

ABDOMINAL AND LUMBAR STRETCHES

RECTUS ABDOMINIS
This muscle originates at the pubis, and inserts at the 5th, 6th, and 7th ribs and the sternum. Its main function is bending the upper body, but it also compresses the abdominal cavity and contributes to the support and protection of the internal organs, as well as to correct posture.

OBLIQUES
There are external and internal obliques.
External oblique: This muscle originates at ribs 5 through 12, and it inserts into the iliac crest, the thoracolumbar fascia, the linea alba abdominis, and the pubis.

Internal oblique: This muscle originates at the iliac crest, the thoracolumbar fascia, and the inguinal ligament, and it inserts at ribs 9 through 12, the aponeurosis of the transverse abdominis muscle, the inguinal ligament, the linea alba abdominis, and the cartilage of ribs 7 though 9.

The main function of both muscles is the rotation of the torso, and they work with the rectus abdominis in bending the upper body, compressing the abdominal cavity, supporting and protecting the internal organs, and maintaining correct posture.

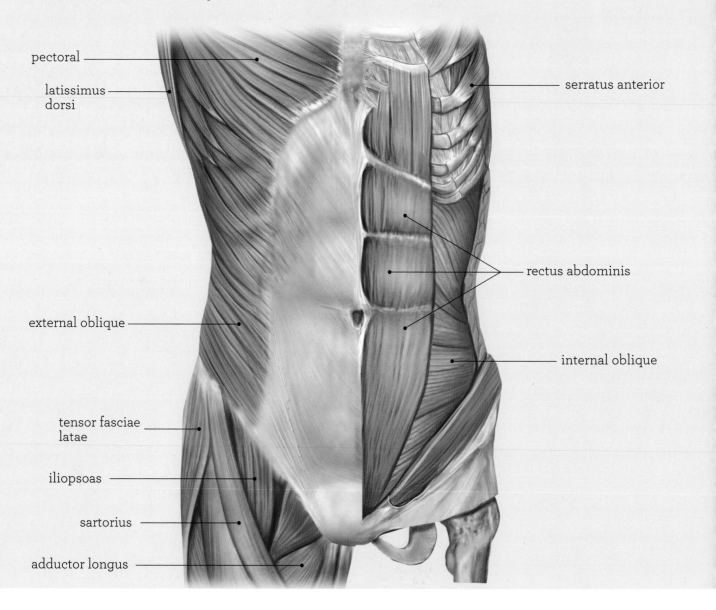

pectoral

latissimus dorsi

serratus anterior

rectus abdominis

external oblique

internal oblique

tensor fasciae latae

iliopsoas

sartorius

adductor longus

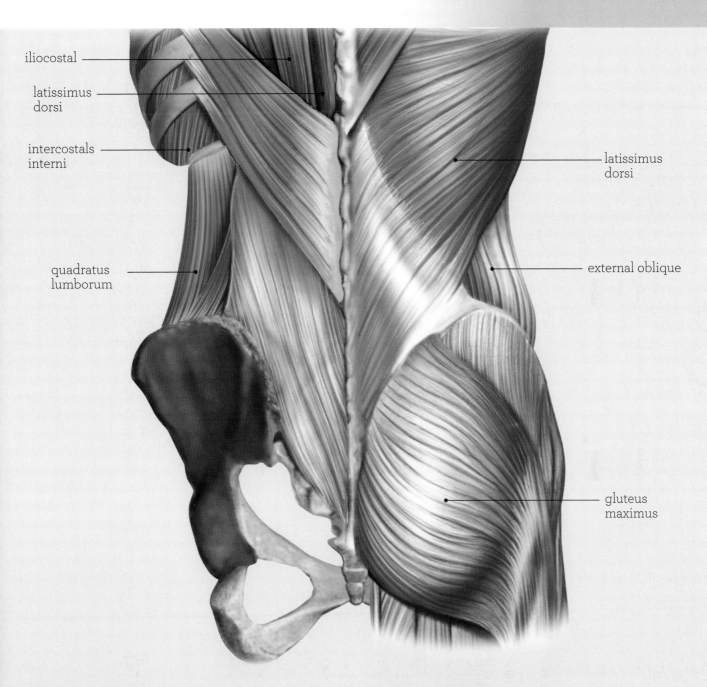

iliocostal

latissimus dorsi

intercostals interni

quadratus lumborum

latissimus dorsi

external oblique

gluteus maximus

QUADRATUS LUMBORUM

This muscle originates at the ilion (crest and inner edge), and it inserts at the lower edge of the 12th rib and at the transverse process of vertebrae L1 to L4. Its main functions are the extension of the lumbar vertebrae and the lateral flexion of the torso. The function of this muscle counters that of the abdominal muscles, so an imbalance between them may entail imbalances in posture and the lumbar vertebrae.

LUMBAR ILIOCOSTAL

The iliocostal is a very long muscle that extends from the cervical vertebrae to the sacral crest. Its lumbar section originates in the iliac and sacral crests, and it inserts in ribs 7 through 12. Its main function is the extension of the lumbar vertebrae, and thus it contributes to performing functions very similar to those of the quadratus lumborum.

Supine Spinal Column Extension

START

Lie down on your back, with your legs stretched out and your hands joined above your head. Keep the backs of your hands facing the top of your head in such a way that your palms face upward, as if you were trying to push against something lying on top of you.

TECHNIQUE

Stretch out your arms while maintaining contact between your hands in such a way that they remain as parallel and close to the floor as possible. You will notice that your lumbar vertebrae come off the floor and the curvature in that area becomes more pronounced. This will stretch your abdominal muscles, and you will feel the tension that the exercise produces.

Starting Position

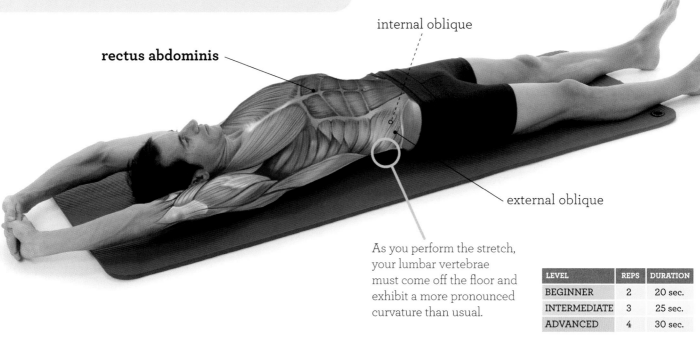

internal oblique

rectus abdominis

external oblique

As you perform the stretch, your lumbar vertebrae must come off the floor and exhibit a more pronounced curvature than usual.

LEVEL	REPS	DURATION
BEGINNER	2	20 sec.
INTERMEDIATE	3	25 sec.
ADVANCED	4	30 sec.

CAUTION

As with all exercises that subject the shoulder to extreme flexion, you will need to take care with your discomfort and reduce the flexion at the slightest pain. The same applies to extending the lumbar vertebrae.

BENEFITS

Reduction of tension in the entire abdominal region, especially the central area.

INDICATION

For rebalancing the body structure in the face of bad posture habits, as well as for relaxing muscle tone after athletic activity.

Kneeling Lumbar Extension

Starting Position

START

Get down on all fours on a mat, keeping your knees aligned with your hips and your hands about shoulder-width apart and slightly to the front, in order to create a stable position.

TECHNIQUE

Extend your spine as if you were trying to throw your abdomen outward, with a movement opposite of what a cat does when it arches its back. You will feel the stretch in your abdominal muscles. Hold it for the time appropriate to your level.

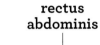

rectus abdominis

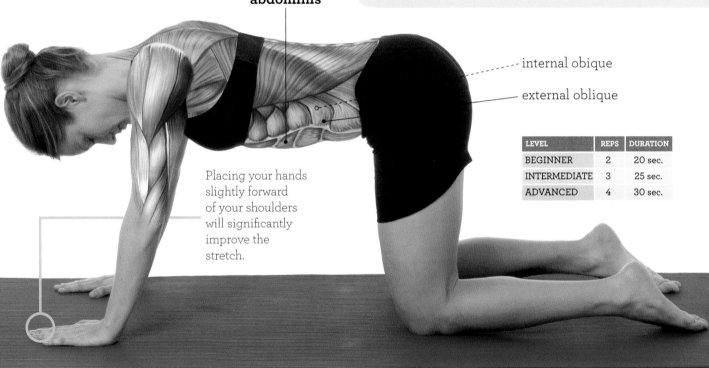

internal obique

external oblique

Placing your hands slightly forward of your shoulders will significantly improve the stretch.

LEVEL	REPS	DURATION
BEGINNER	2	20 sec.
INTERMEDIATE	3	25 sec.
ADVANCED	4	30 sec.

CAUTION

If you have problems in the lumbar area, you will have to be particularly careful as you extend your vertebrae. Stop the movement at any sensation of pain.

BENEFITS

Reduction of tension in the abdominal region, especially the central area.

INDICATION

To relax muscle tone after athletic activity and rebalance your body structure after exhibiting bad posture.

Cobra Position

Starting
Position

START

Lie down on your stomach with your hands pressed against the floor, close to your chest, in a position similar to the starting position for a push-up. Your legs must be extended and your ankles in plantar flexion.

TECHNIQUE

Slowly and gradually straighten your elbows, but keep your upper body relaxed. You must not push yourself up all at once, as in a push-up; instead, it's your chest that moves off the floor while your hips and your legs remain in contact with it. Push yourself up until your hips begin to separate from the floor. At that point, hold the movement, feel the tension in your abdominal wall, and hold the position for the time appropriate to your level.

Remember to keep your back in extension and not lift up all at once.

internal oblique

rectus abdominis

external oblique

LEVEL	REPS	DURATION
BEGINNER	2	15 sec.
INTERMEDIATE	3	20 sec.
ADVANCED	4	30 sec.

CAUTION

Your hips should barely rise a finger's width from the floor, or not at all; otherwise, you have straightened your elbows and raised yourself too much. Be alert to possible discomfort in the lumbar region and stop the movement if it is present.

BENEFITS

Reduction of tension in the entire abdominal region, especially the central area.

INDICATION

For anyone who wants to relax muscle tone after an athletic activity, such as a running race. Also, to rebalance the body structure in the face of bad positions adopted in a daily routine.

Rotation Stretch

Starting Position

LEVEL	REPS	DURATION
BEGINNER	2	20 sec.
INTERMEDIATE	4	25 sec.
ADVANCED	5	35 sec.

START
Lie down on your back on a mat. Your arms need to go out to the sides, and the arm on the side to be stretched should form a 90° angle with your upper body. The leg on this same side needs to be bent so that the sole of your foot contacts the floor. The other leg is totally straight and in line with your upper body.

TECHNIQUE
Rotate your upper body so that the upper part of your back remains in contact with the floor, but the lower part gradually separates from it. Cross your bent leg over the other one, as if you were trying to touch the floor with the inside of your knee, but without doing so.

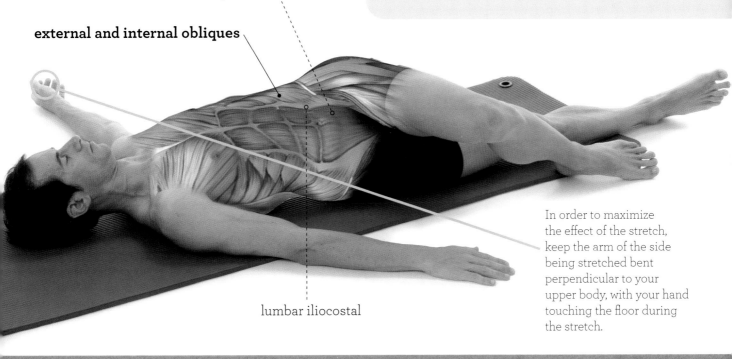

quadratus lumborum

external and internal obliques

In order to maximize the effect of the stretch, keep the arm of the side being stretched bent perpendicular to your upper body, with your hand touching the floor during the stretch.

lumbar iliocostal

CAUTION
This stretch involves a pronounced rotation of the torso, which may cause discomfort in people with prior problems with their spines. If this applies to you, do the stretch very slowly and gradually, and be attentive to any discomfort.

BENEFITS
Reduction of tension in the abdominal and lumbar region, especially in the side area.

INDICATION
For relaxing muscle tone after athletic activity, as well as for rebalancing a body position altered by poor posture.

Forearms on the Head

START

Stand with your legs straight and your feet aligned with your shoulders. Then raise your arms and bend your elbows so that your forearms are on top of your head and you can grasp them with your hands.

TECHNIQUE

Rotate your upper body in such a way that your shoulders are no longer aligned with your hips. You will feel tension in your abdominal area on the side opposite the direction in which you have turned. Hold the tension for a few seconds and return to the starting point. Pause before doing the exercise again.

deep back muscles

external oblique

internal oblique

Starting Position

Keep your feet far enough apart to assure stability during the entire exercise.

LEVEL	REPS	DURATION
BEGINNER	2	20 sec.
INTERMEDIATE	4	25 sec.
ADVANCED	5	35 sec.

CAUTION

Be sure to start from a stable position and to keep your balance, because rotating your torso may contribute to instability.

BENEFITS

Reduction of tension in the abdominal region, especially in the side.

INDICATION

To relax muscle tone after athletic activity and to contribute to improving your posture.

Upper Body Lateral Lean and Bend

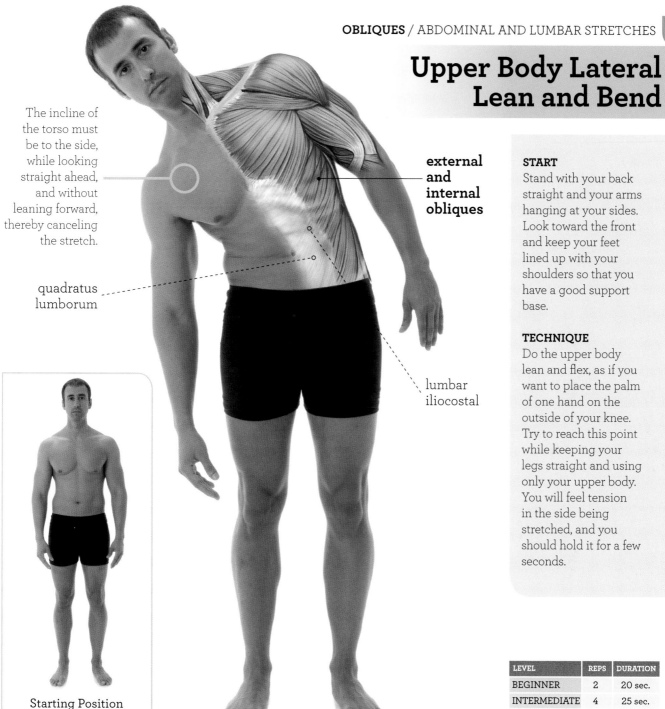

The incline of the torso must be to the side, while looking straight ahead, and without leaning forward, thereby canceling the stretch.

quadratus lumborum

external and internal obliques

lumbar iliocostal

Starting Position

START
Stand with your back straight and your arms hanging at your sides. Look toward the front and keep your feet lined up with your shoulders so that you have a good support base.

TECHNIQUE
Do the upper body lean and flex, as if you want to place the palm of one hand on the outside of your knee. Try to reach this point while keeping your legs straight and using only your upper body. You will feel tension in the side being stretched, and you should hold it for a few seconds.

LEVEL	REPS	DURATION
BEGINNER	2	20 sec.
INTERMEDIATE	4	25 sec.
ADVANCED	5	35 sec.

CAUTION	BENEFITS	INDICATION
It is important to create a good support base to maintain your balance.	Reduction of tension in the sides of your upper body.	For preventing tension caused by poor posture.

Stretch with Knees Against Chest

START

Lie down on your back on a mat and hold your legs by the knees. Keep your head on the floor and your back straight.

TECHNIQUE

Pull your knees toward your chest to produce a rearward tilt in your hips, while keeping your lumbar vertebrae in contact with the mat and canceling out the natural curvature of your spine in this area. Feel the stretch in the lower part of your back and hold it for a few seconds.

Starting Postion

LEVEL	REPS	DURATION
BEGINNER	3	20 sec.
INTERMEDIATE	4	30 sec.
ADVANCED	5	45 sec.

gluteus maximus

Keep the lower part of your back in contact with the mat during the stretch.

quadratus lumborum

lumbar iliocostal

latissimus dorsi

CAUTION

It is common to exert tension in the neck when you pull on your knees. Avoid subjecting your neck to this tension, which may be harmful, and let your head rest on the mat.

BENEFITS

Reduction of tension in the entire lumbar region and correction of posture.

INDICATION

To avoid shortening of the lumbar muscles and reduce pain in people who suffer from lower back problems. Also, to improve posture in people with lumbar hyperlordosis and reduce the sense of lumbar strain in people who spend lots of time in the same position, whether standing or sitting.

Crossed Leg

Starting Position

START

Sit on a mat with one leg stretched out and the other knee bent and crossed over it. Also hold the arm corresponding to the leg being stretched as shown in the starting photo, and rest the other hand on the mat to provide support.

TECHNIQUE

Drop the crossed arm to the outside of the bent leg and use it to turn your upper body toward the side of your support hand. This rotation brings your shoulders and hips out of alignment, and produces a stretch in your lower back muscles.

Rotate your upper body.

LEVEL	REPS	DURATION
BEGINNER	3	20 sec.
INTERMEDIATE	4	30 sec.
ADVANCED	5	45 sec.

iliocostal

deep back muscles

quadratus lumborum

gluteus maximus
gluteus medius
gluteus minimus

CAUTION	BENEFITS	INDICATION
Concentrate on the rotation of the torso and not on pulling on the bent leg, because this is the difference between stretching the back muscles and stretching the gluteals.	Reduction of tension in the lumbar region.	For people with lower back problems or lumbar hyperlordosis. It reduces the sensation of lumbar strain in individuals who spend lots of time standing or sitting.

Knee Bend onto Chest

START
Lie down on your back with your head touching the mat. Keep one leg stretched out in line with your upper body and raise the other one while bending your knee and holding it with your hands.

TECHNIQUE
Pull your knee toward your chest while keeping the other leg stretched out and parallel to the floor. You will first notice tension in the lower part of your back, on the side of the bent leg. Hold this tension for a few seconds.

Starting Position

Keep your head in contact with the mat to avoid tension in your neck.

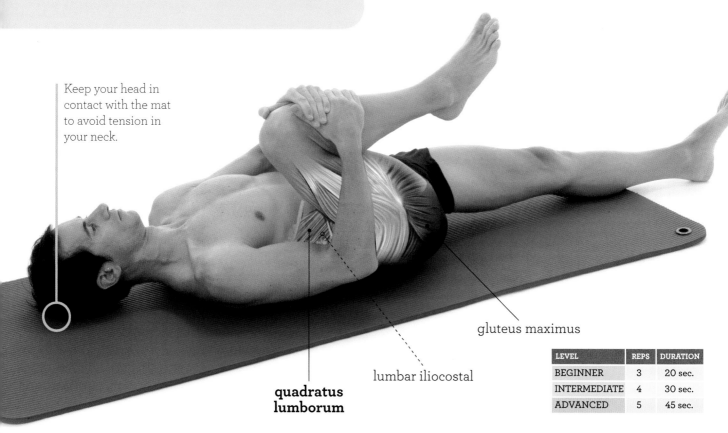

gluteus maximus

lumbar iliocostal

quadratus lumborum

LEVEL	REPS	DURATION
BEGINNER	3	20 sec.
INTERMEDIATE	4	30 sec.
ADVANCED	5	45 sec.

CAUTION
Even though it will be difficult to keep your straight leg totally in contact with the mat, try to keep it as close as possible and parallel to it.

BENEFITS
Reduction of tension in the lumbar region.

INDICATION
For people with lower back problems or lumbar hyperlordosis. This exercise reduces the sensation of lumbar strain in individuals who spend lots of time standing or sitting.

Seated Upper Body Bend

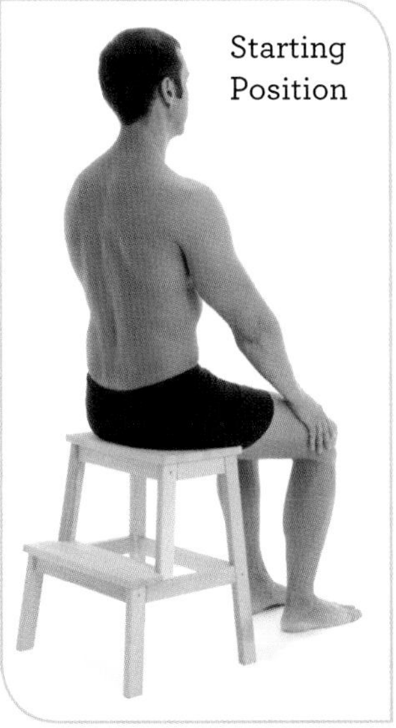

Starting Position

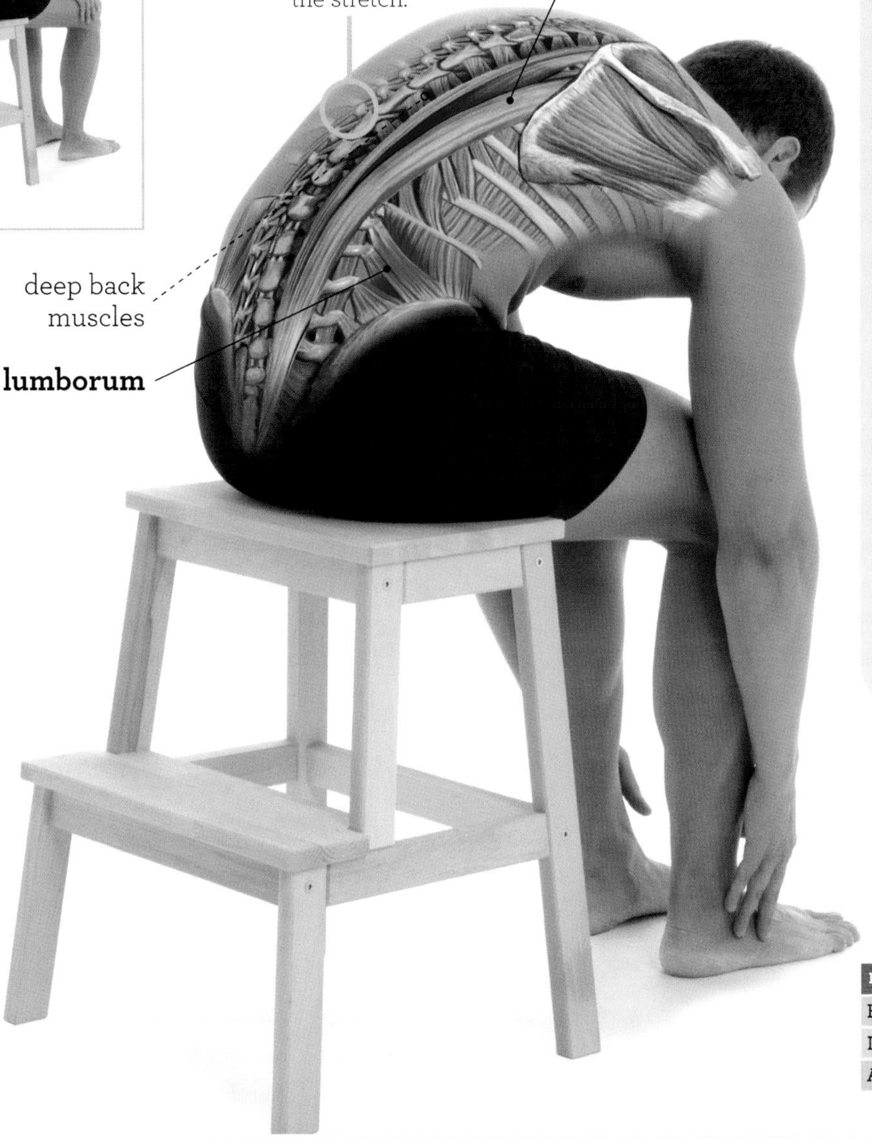

Lean and bend your spine during the stretch.

iliocostal

deep back muscles

quadratus lumborum

START
Sit on a chair or a stool with your back straight and your hands on your knees. Look to the front and keep your feet on the floor.

TECHNIQUE
Bend your upper body forward as you slide your hands toward your ankles. Try to keep your hips turned rearward, as if you were trying to keep the middle and rear of your gluteals in contact with the seat. Feel the stretch of your back muscles and hold the position for a few seconds.

LEVEL	REPS	DURATION
BEGINNER	3	20 sec.
INTERMEDIATE	4	30 sec.
ADVANCED	5	45 sec.

CAUTION

Avoid rolling your hips toward the front, because that will reduce the effect of the stretch on the back muscles.

BENEFITS

Reduction of tension in the back muscles.

INDICATION

For people who spend lots of time standing, especially in front of a counter, or who lift and move heavy objects.

Crouching Upper Body Bend

START
Crouch down with your arms inside your legs, your elbows bent, and your hands clasped. Your upper body will lean forward.

TECHNIQUE
Bend your upper body and lean forward so you feel the tension along your back. At the point of maximum tension, be sure to keep your balance and hold the position for the appropriate time for your level.

At the point of maximum tension, also bend your neck and head to increase the stretch of the back muscles.

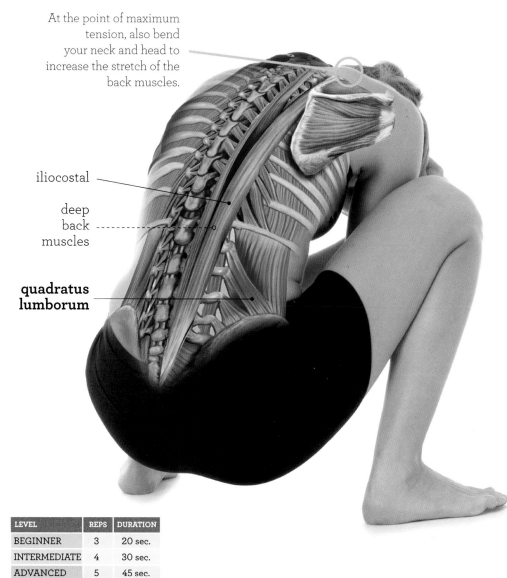

iliocostal

deep back muscles

quadratus lumborum

Starting Position

LEVEL	REPS	DURATION
BEGINNER	3	20 sec.
INTERMEDIATE	4	30 sec.
ADVANCED	5	45 sec.

CAUTION

Start from a stable, well-balanced position, because the forward lean of the torso can throw you off balance.

BENEFITS

Reduction of tension in the back muscles.

INDICATION

For people who spend many hours standing or who lift and move heavy objects.

Seated with Arms Forward

Starting Position

LEVEL	REPS	DURATION
BEGINNER	3	20 sec.
INTERMEDIATE	4	30 sec.
ADVANCED	5	45 sec.

START
Sit on a mat with your legs together and stretched out, and extend your arms forward. Look to the front and keep your back straight.

TECHNIQUE
Bend your upper body and try to touch your feet with your fingers. Try to keep your hips from rolling to the front, as if you were trying to keep the middle and back of your gluteals in contact with the floor, as in stretch number 27. You can use a slight bend in the knees if you feel too much tension in the back of your legs.

spinal erectors

Keep your hips from rolling forward.

quadratus lumborum

ischiotibials

CAUTION	BENEFITS	INDICATION
Do the stretch gradually to make sure that the tension is produced in the back muscles.	Reduction of tension in the entire lumbar area.	For people with strain in their back muscles, especially if it is due to positions held for many hours during the day.

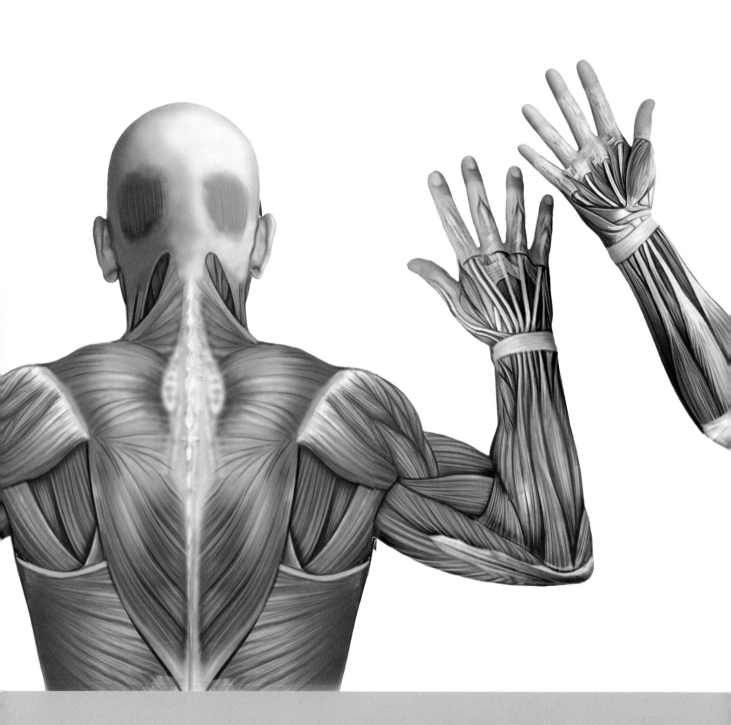

UPPER EXTREMITIES,
SHOULDERS, AND CHEST STRETCHES

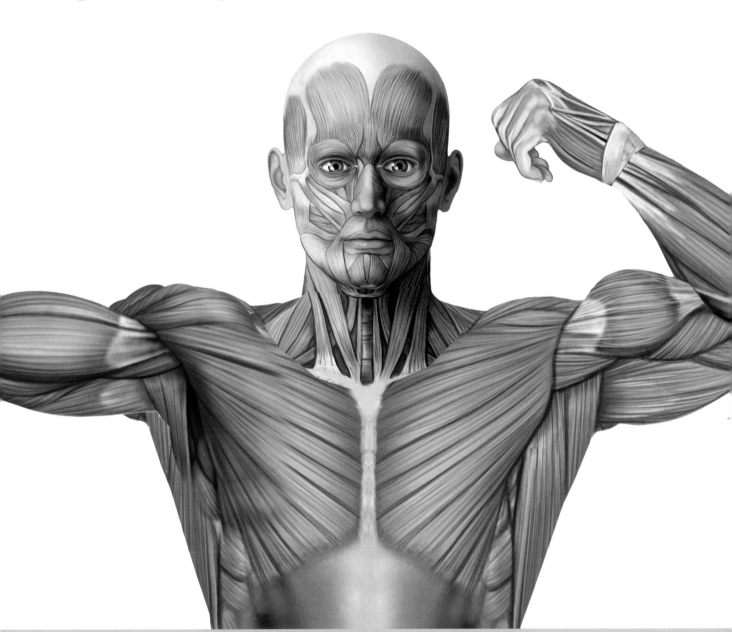

SHOULDER STRETCHES

DELTOID
Even though this is a single muscle, its distinct sections perform different functions.

Anterior section: It originates at the distal third of the clavicle, and it inserts at the deltoid process of the humerus, where it coincides with the two remaining sections. Its main function is the antepulsion or flexion of the shoulder, the movement that results when raising the arm toward the front.

Medial section: It originates at the scapula (spinal crest and acromion), and it inserts at the deltoid tuberosity of the humerus. Its main function is shoulder abduction.

Posterior section: The origin is at the spine of the scapula, and its main function is the extension or retropulsion of the shoulder, a function opposite that of the preceding section.

PECTORALIS MAJOR
This muscle originates at the anterior surface of the clavicle, at the body of the sternum, in the anterior costal cartilages of the first six ribs, and at the process of the oblique muscle, and it inserts at the intertubercular groove of the humerus. Its functions are the flexion, adduction, and inward rotation of the shoulder. Given the extent of its origin, it is possible to do flexibility and strength exercises from many different angles, and all of them are effective.

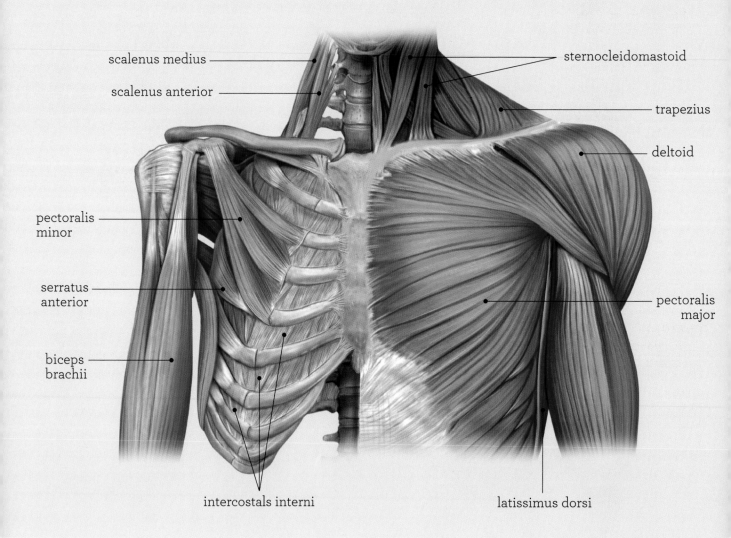

scalenus medius

scalenus anterior

pectoralis minor

serratus anterior

biceps brachii

intercostals interni

sternocleidomastoid

trapezius

deltoid

pectoralis major

latissimus dorsi

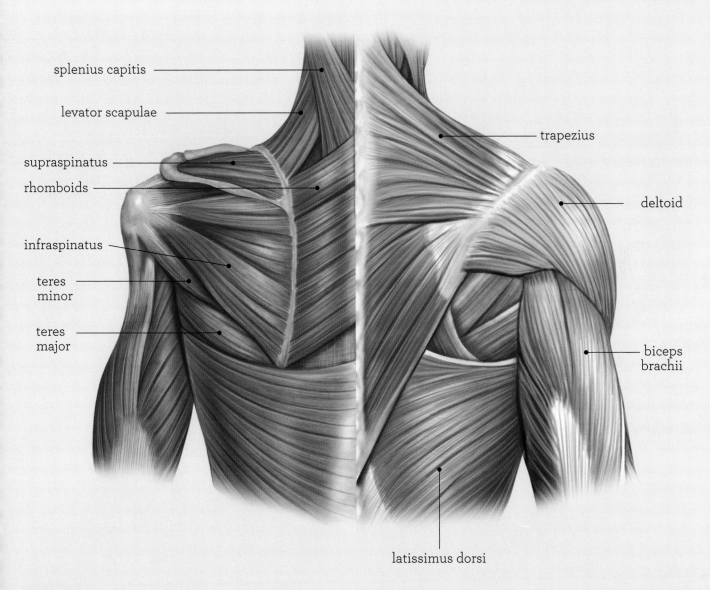

splenius capitis

levator scapulae

supraspinatus

rhomboids

infraspinatus

teres
minor

teres
major

trapezius

deltoid

biceps
brachii

latissimus dorsi

ROTATORS

The shoulder can rotate inward and outward. In both cases, different muscles are used, and we will describe the most important ones. The muscles that are used in rotating the shoulder, such as the pectoralis major and the latissimus dorsi, are also described in other sections of the book.

Infraspinatus: This muscle originates at the infraspinous fossa of the scapula, and it inserts at the tuberculum majus humeri. Its main function is the outward rotation of the shoulder, although it is used also for shoulder abduction.

Teres minor: This muscle originates at the lateral edge of the scapula, and it inserts at the tuberculum majus humeri. Its main function is outward rotation of the shoulder.

Subscapularis: This muscle arises at the subscapular fossa, and it inserts at the tuberculum minus humeri. Its main function is inward rotation of the shoulder.

Teres major: This muscle arises at the lower angle of the scapula, and it inserts at the intertubercular groove of the humerus. Its main function is inward rotation of the shoulder, but it also participates in shoulder adduction and extension.

Posterior with Arm in Front

START

Stand and cross one arm straight in front of your chest. Place the other forearm over the crossed arm and block it in place. Remember to keep your back straight, look toward the front, keep an adequate support base, and maintain symmetry in both legs.

TECHNIQUE

Pull the straight arm as tight as possible to your chest using the other forearm. At the point where you reach the maximum tension, hold the position to prolong the stretch. As with some other stretches, it may be more difficult to feel the tension or to identify the stretching sensation than in an exercise for the ischiotibials, but this does not mean that the exercise is not being done properly.

teres minor

deltoid posterior

infraspinatus

Keep your upper body straight, without rotating it.

Starting Position

LEVEL	REPS	DURATION
BEGINNER	3	20 sec.
INTERMEDIATE	3	35 sec.
ADVANCED	4	50 sec.

CAUTION

Rotating the torso during the stretch may give a false sense of achieving a greater range of stretching, but in reality this adds nothing, so it should be avoided.

BENEFITS

Reduction of tension in the back and side area of the arm, and an increase in the range of movement in the shoulder joint.

INDICATION

For people who lift heavy weights or do repetitive movements with their shoulders, as well as for people who do certain sports: swimming, baseball, tennis, and golf.

Posterior with Anchor Point

teres minor

posterior deltoid

infraspinatus

The anchor point should be about shoulder height.

Starting Position

START

Stand next to some object you can hold onto, whether it's a corner, a door frame, or any other solid point. Take a position next to it and hold onto it with the opposite hand; you will need to rotate your upper body to do this. Keep your feet apart to assure a good support base.

TECHNIQUE

Try to line up your shoulders with your feet, as if you were trying to face straight ahead once again, but without letting go with the hand holding the fixed point. You will feel the tension from the stretch in the rear of your shoulder. Hold the position for a few seconds.

LEVEL	REPS	DURATION
BEGINNER	3	20 sec.
INTERMEDIATE	3	35 sec.
ADVANCED	4	50 sec.

CAUTION

Avoid moving your feet during the exercise, especially when you hold onto the anchor point. The distance between the position of your shoulders and that of your feet can then be reduced in order to do the stretch properly.

BENEFITS

Reduction of tension in the rear and side of the arm, and an increase in the range of movement in the shoulder joint.

INDICATION

For people whose work requires them to carry heavy weights or perform repetitive movements with their shoulders, as well as for athletes whose discipline involves moving the shoulder repeatedly, as in golf, swimming, and racquet sports.

Anterior with Arms Behind

START

Stand with your back straight and look straight ahead. Clasp your hands behind your back and keep your feet shoulder-width apart to create a balanced position.

TECHNIQUE

Try to raise your hands toward the rear through retropulsion of your shoulders, until you reach a point where you cannot raise them any farther. Hold the position and maintain the movement while continually pulling upward.

anterior deltoid

pectoralis major

coracobrachialis

Keep your hands clasped.

Starting Position

LEVEL	REPS	DURATION
BEGINNER	3	20 sec.
INTERMEDIATE	4	30 sec.
ADVANCED	5	35 sec.

CAUTION

Avoid leaning your torso in order to raise your hands. Remember to stay perpendicular to the floor while performing the stretch.

BENEFITS

Reduction of tension in the front and side of your arm and optimization of the range of movement in the shoulder joint.

INDICATION

For people who often do sports using implements and swimming, as well as for people whose usual activity involves pushing heavy objects: wheelchairs, stretchers, carts, etc.

Starting Position

Anterior While Seated

LEVEL	REPS	DURATION
BEGINNER	3	20 sec.
INTERMEDIATE	4	30 sec.
ADVANCED	5	35 sec.

START

Sit on a mat with your upper body slightly inclined to the rear and support yourself on both hands slightly behind your shoulders. Your feet will contact the mat and your knees will be bent at 90°.

TECHNIQUE

Slide your buttocks toward your feet, as if you were trying to sit closer to them, and keep your hands anchored at the initial support points. You will notice the muscle tension in the front of your shoulders. Maintain it for the time appropriate to your level.

Keep your hands anchored on the same points during the entire exercise.

anterior deltoid

coracobrachial

pectoralis major

CAUTION

Do the exercise slowly and gradually to avoid forcing your shoulders and elbows. If you have back problems, skip this exercise and choose a different one from the same muscle group.

BENEFITS

Reduction of tension in the front and side of the arm, and optimization of the range of movement in the shoulder joint.

INDICATION

For people who must regularly push carts, wheelchairs, or heavy objects, and for people who regularly do sports with implements, especially racquets.

Bilateral with Upper Body Bend and Wall Support

START
Stand in front of a wall, with both hands contacting it above shoulder level.

TECHNIQUE
Bend your upper body and lower your chest and shoulders until you feel the tension in your pectorals. Hold the position for the appropriate time to prolong the stretch.

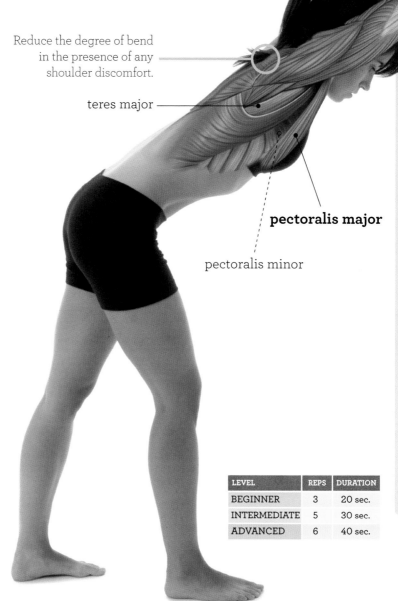

Reduce the degree of bend in the presence of any shoulder discomfort.

teres major

pectoralis major

pectoralis minor

Starting Position

LEVEL	REPS	DURATION
BEGINNER	3	20 sec.
INTERMEDIATE	5	30 sec.
ADVANCED	6	40 sec.

CAUTION

This exercises produces a very pronounced antepulsion or flexion, so you need to be very careful and attentive to the slightest sensation of pain that it produces. If you feel pain, significantly reduce the shoulder antepulsion.

BENEFITS

Reduction of tension in the front of the shoulder, optimization of its range of movement, improved posture, and correction of kyphosis.

INDICATION

For strength and bodybuilding athletes, and for those who do cyclical shoulder movements, such as swimmers. Also for people who tend to slouch, who spend many hours seated at a desk or a computer, who perform repetitive shoulder movements, or who carry heavy weights at work.

Unilateral Wall Support

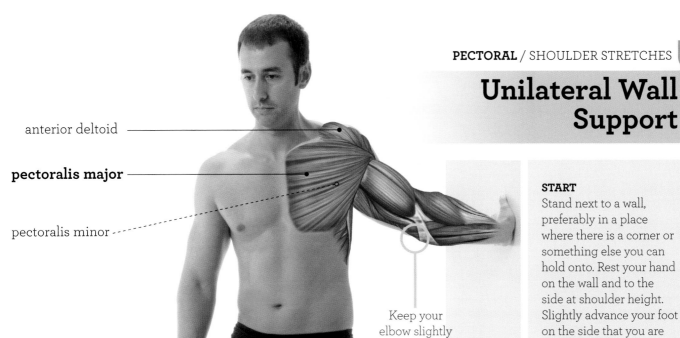

anterior deltoid

pectoralis major

pectoralis minor

Keep your elbow slightly bent.

Starting Position

START

Stand next to a wall, preferably in a place where there is a corner or something else you can hold onto. Rest your hand on the wall and to the side at shoulder height. Slightly advance your foot on the side that you are going to stretch.

TECHNIQUE

Rotate your upper body, as if you wanted to turn your back to the wall, but without removing your hand from it or moving your feet. When you feel the stretch in your pectoral, you have reached the right position. Hold the tension.

LEVEL	REPS	DURATION
BEGINNER	3	20 sec.
INTERMEDIATE	5	30 sec.
ADVANCED	6	40 sec.

CAUTION

Keep a slight bend in the elbow of the support arm to be sure that you stretch the pectoral and not the biceps brachii.

BENEFITS

Stretching the pectoral and correcting kyphosis, mainly when combined with exercises for strengthening the back.

INDICATION

For people with imbalances between the dorsal and pectoral muscles, which generally lead to muscle spasms, back pains, and bad posture.

Support with Bent Elbow

anterior deltoid ———————

pectoralis major ———————

pectoralis minor ----------

START
Stand next to a wall, a trellis, or some similar support structure. Stand sideways to this anchor point and with the foot closer to it slightly ahead of the other. Rest your forearm on the wall, with your hand upward and your elbow about shoulder height.

TECHNIQUE
Turn your upper body away from the wall while keeping your forearm in contact with it. As the rotation of the upper body progresses, you will feel tension in your chest, which indicates that you are doing the stretch correctly. Hold the position for a few seconds without reducing the tension, and then return to the starting point.

Starting Position

LEVEL	REPS	DURATION
BEGINNER	3	20 sec.
INTERMEDIATE	4	35 sec.
ADVANCED	6	45 sec.

Move the foot closer to the support point ahead slightly.

CAUTION
Reduce the tension of the stretch if you feel any discomfort in the stretch.

BENEFITS
Reduction of stress in the front part of the shoulder, wider range of shoulder movement, improved posture, and correction of kyphosis.

INDICATION
For people with imbalances between the muscles in the back and the pectoral area, and strength and hypertrophy athletes. Also for people who regularly push heavy objects, such as wheelchairs, carts, stretches, and other similar items.

Rear Arm Extension

anterior deltoid

coracobrachialis

pectoralis major

Keep your upper body perpendicular to the floor.

Starting Position

START
Stand with your back straight and your feet shoulder-width apart. Hold a stick behind your back with both hands palms down. If you don't have a stick, you can do the stretch while holding onto a trellis, or with both hands resting on a piece of furniture, such as a table or a desk.

TECHNIQUE
Raise the stick behind you while keeping your upper body completely perpendicular to the floor. You will feel tension in your pectorals. Hold the stretch for a few seconds before returning to the starting point. If you are doing the stretch without a stick, and using an anchor point instead, you will have to gradually bend your knees so that your upper body lowers until you feel the muscle tension.

LEVEL	REPS	DURATION
BEGINNER	3	20 sec.
INTERMEDIATE	4	35 sec.
ADVANCED	5	45 sec.

CAUTION
Pay particular attention to the sensations in your shoulders and stop if you feel any discomfort. If you choose to do this exercise by holding onto a fixed point, keep your elbows slightly bent to focus the stretch in the pectorals and not in the biceps.

BENEFITS
Relief of tension in the front part of the shoulder, greater range of movement in the shoulder, correct posture, and avoidance of kyphosis.

INDICATION
For people with kyphosis due to muscle imbalances, people who habitually push heavy objects, and strength and hypertrophy athletes.

Hands on Head

START

Stand with both hands on the back of your head and your elbows forward. Keep your back straight and your feet shoulder-width apart. Avoid pulling against your head.

TECHNIQUE

Gradually move your elbows apart and to the rear until they won't go back any farther. You will feel tension in your pectorals. Hold the stretch for a few seconds before returning to the starting position.

pectoralis minor

pectoralis major

serratus anterior

Place your hands on the back of your head without pulling.

Starting Position

LEVEL	REPS	DURATION
BEGINNER	3	20 sec.
INTERMEDIATE	4	35 sec.
ADVANCED	5	45 sec.

CAUTION

Avoid pulling against your head and subjecting the cervical vertebrae to unnecessary tension.

BENEFITS

Reduction of tension in the front part of the shoulder, improved posture, and correction of kyphosis.

INDICATION

For people with imbalances between the dorsal and pectoral areas, people who regularly push heavy weights, such as carts, wheelchairs, and similar items, and strength and hypertrophy athletes.

Seated with Hands on Chest

Starting Position

trapezius

rhomboids

infraspinatus

teres minor

latissimus dorsi

Keep your thumbs in your armpits so your hands stay in the right position.

START
Sit on a mat, with your feet resting on it and your knees bent at around 90°. Support your elbows on the inside of your knees and place your hands on the top of your chest with your thumbs in your armpits.

TECHNIQUE
Move your knees together so that they push your elbows inward, increasing the tension on the external rotators of the shoulder. Hold the tension for a few seconds before returning to the starting point.

LEVEL	REPS	DURATION
BEGINNER	2	20 sec.
INTERMEDIATE	3	25 sec.
ADVANCED	4	30 sec.

CAUTION
Be sure to place your hands correctly so that the stretch is not lost part of the way through the exercise.

BENEFITS
Stretching the external rotators of the shoulder, with the resulting decrease of tension in this region.

INDICATION
For people who do repetitive shoulder movements and people whose work requires carrying heavy weights.

Forward Elbow Pull

START

Stand and place one hand on your hip on the same side, such that your elbow is bent around 90°. Using the opposite hand, grasp the bent elbow and keep your back straight.

TECHNIQUE

Pull your elbow forward slowly and gradually, keeping in mind that the movement will be short and you should not force your shoulder. The hand on your waist must not move from its anchor point at any time.

infraspinatus ———  ——— rhomboid

teres minor

Pull your elbow forward.

Starting Position

LEVEL	REPS	DURATION
BEGINNER	2	20 sec.
INTERMEDIATE	3	25 sec.
ADVANCED	4	30 sec.

CAUTION

Perform the movement slowly and stop the stretch if you feel discomfort in the shoulder.

BENEFITS

Reduction of tension in the rear of the shoulder.

INDICATION

For people who do repetitive shoulder movements or regularly lift heavy weights using their arms and hands.

Crank Position

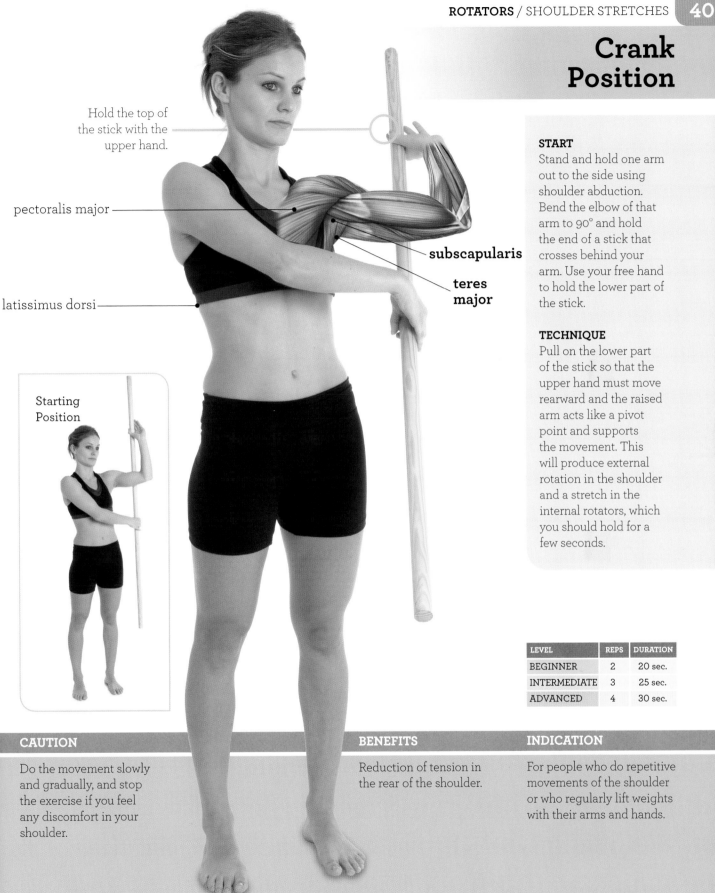

Hold the top of the stick with the upper hand.

pectoralis major

subscapularis

teres major

latissimus dorsi

Starting Position

START
Stand and hold one arm out to the side using shoulder abduction. Bend the elbow of that arm to 90° and hold the end of a stick that crosses behind your arm. Use your free hand to hold the lower part of the stick.

TECHNIQUE
Pull on the lower part of the stick so that the upper hand must move rearward and the raised arm acts like a pivot point and supports the movement. This will produce external rotation in the shoulder and a stretch in the internal rotators, which you should hold for a few seconds.

LEVEL	REPS	DURATION
BEGINNER	2	20 sec.
INTERMEDIATE	3	25 sec.
ADVANCED	4	30 sec.

CAUTION
Do the movement slowly and gradually, and stop the exercise if you feel any discomfort in your shoulder.

BENEFITS
Reduction of tension in the rear of the shoulder.

INDICATION
For people who do repetitive movements of the shoulder or who regularly lift weights with their arms and hands.

ARM AND FOREARM STRETCHES

BICEPS BRACHII

This muscle consists of two parts. The first originates at the coracoid process of the scapula, and it inserts at the radial tuberosity. The second one arises at the supraglenoid tuberculum of the scapula, and it inserts at the bicipital aponeurosis.

TRICEPS BRACHII

This muscle is made up of three parts that originate at the infraglenoid process of the scapula and the diaphysis of the humerus. The three parts have a common insertion at the olecranon of the ulna.

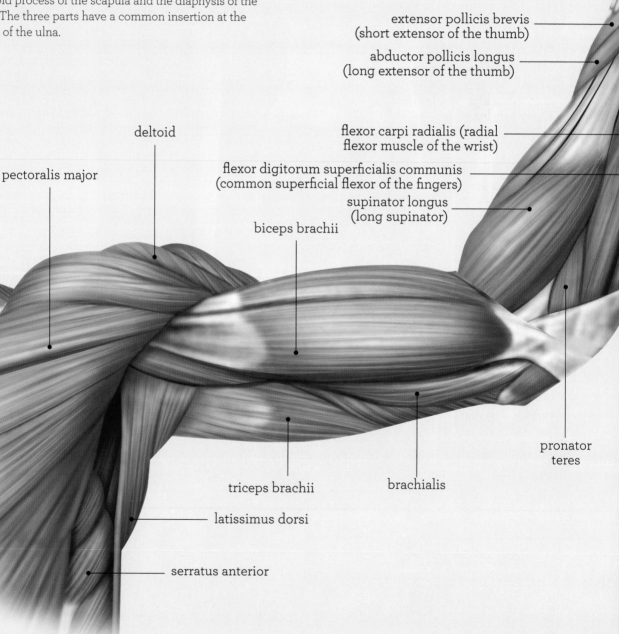

extensor pollicis brevis
(short extensor of the thumb)

abductor pollicis longus
(long extensor of the thumb)

deltoid

flexor carpi radialis (radial
flexor muscle of the wrist)

pectoralis major

flexor digitorum superficialis communis
(common superficial flexor of the fingers)

supinator longus
(long supinator)

biceps brachii

pronator
teres

triceps brachii

brachialis

latissimus dorsi

serratus anterior

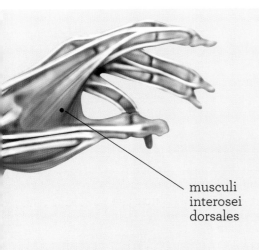

musculi
interosei
dorsales

flexor carpi ulnaris
(ulnar flexor of the wrist)

EPITROCHLEAR MUSCLES

These muscles are located in the forearm, and their main function is bending the wrist. The most significant muscles are the palmaris longus and the flexor carpi ulnaris.

Palmaris longus: This muscle originates at the medial epicondyle of the humerus, and it inserts at the second and third metacarpals.

Flexor carpi ulnaris: This muscle has two parts. It originates at the medial epicondyle of the humerus and the olecranon of the ulna. Its insertion is at the pisiform bone, the hamate bone, and the fifth metacarpal.

EPICONDYLIAN MUSCLES

The main function of these muscles is the extension of the wrist. The most important muscles are the brachoradialis and the extensor carpi radialis brevis.

Extensor carpi radialis longus: This muscle originates at the supracondylar crest of the humerus and inserts at the second metacarpal.

Extensor carpi radialis brevis: This muscle originates at the epicondyle of the humerus, and it inserts at the third metacarpal.

muscili interosei dorsales manus
(dorsal interosseous muscles of
the hand)

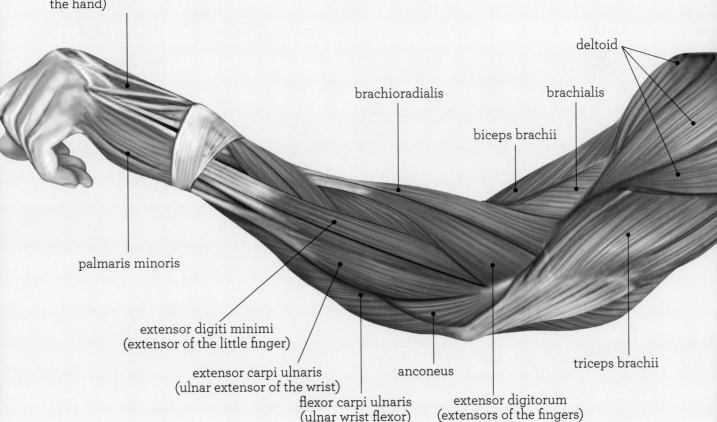

deltoid

brachioradialis

brachialis

biceps brachii

palmaris minoris

extensor digiti minimi
(extensor of the little finger)

extensor carpi ulnaris
(ulnar extensor of the wrist)

flexor carpi ulnaris
(ulnar wrist flexor)

anconeus

extensor digitorum
(extensors of the fingers)

triceps brachii

Inverted Unilateral Support

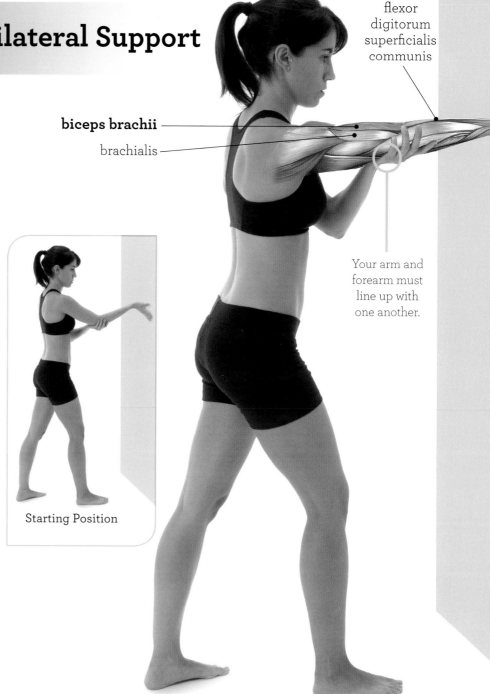

flexor
digitorum
superficialis
communis

biceps brachii —————

brachialis —————

Your arm and
forearm must
line up with
one another.

Starting Position

START
Stand in front of a
wall or some similar
support surface that is
perpendicular to the
floor. Anchor the palm
of your hand with your
fingers facing downward,
at about chest height. In
order to achieve a better
support base and give
a little push, move one
foot forward.

TECHNIQUE
Completely straighten
the elbow of the support
arm such that your
arm and forearm are in
line with one another.
At this point, you will
feel the tension in the
front of your arm and
your elbow, a sign that
your biceps muscle is
stretching properly.

LEVEL	REPS	DURATION
BEGINNER	2	20 sec.
INTERMEDIATE	3	25 sec.
ADVANCED	4	35 sec.

CAUTION
When you straighten your elbow,
if the angle goes much beyond
180°, the range of this joint and the
stretching capacity of your biceps is
greater than normal, so you should
not insist on stretching. Otherwise,
you could endanger the integrity of
your elbow joint.

BENEFITS
Relaxation in the front of
your arm and prevention of
shortening of the biceps.

INDICATION
For people who do repetitive
bending movements of the
elbow, as well as for those
whose work often requires
lifting heavy weights with
their arms and hands.

Wall Support with a Twist

pectoralis major

supinator longus

brachialis

biceps brachii

Keep the support hand anchored on one point during the entire exercise.

START
Stand with your side toward a wall or other support point. Rest the palm of your hand on a point slightly behind your upper body and a little below shoulder height. The foot closer to the wall should be placed ahead of the other one.

TECHNIQUE
Straighten your elbow without moving your hand, and slightly twist the upper part of your shoulders away from the support point, as if you wanted to turn your back to it. You will feel the tension in the front part of your elbow. Hold for a few seconds before returning to the starting point.

Starting Position

LEVEL	REPS	DURATION
BEGINNER	2	30 sec.
INTERMEDIATE	3	25 sec.
ADVANCED	4	35 sec.

CAUTION
Remember to straighten your elbow when you do the stretch. Otherwise, this exercise would affect the pectoral more than the biceps. Skip this exercise if your brachial plexus is affected by any disorder or injury.

BENEFITS
Reduction of tension in the front part of the arm and maintaining a proper range of movement in the elbow.

INDICATION
For people who do repetitive elbow bends, for those whose work makes them carry heavy weights with their arms and hands, and for strength and hypertrophy athletes.

Pulling from Behind

START

Stand with your back to a fixed point no higher than your shoulders and no lower than your waist. Reach rearward with one hand and hold onto the fixed point with your palm facing inward. One foot needs to be slightly ahead of the other.

TECHNIQUE

Without letting go with the support hand, begin to bend both knees slightly. You will feel a gradual increase in tension in the front part of your arm and elbow. When you begin to feel discomfort, and before it becomes painful, stop the movement and hold it for a few seconds before returning to the starting position.

LEVEL	REPS	DURATION
BEGINNER	2	20 sec.
INTERMEDIATE	3	25 sec.
ADVANCED	4	35 sec.

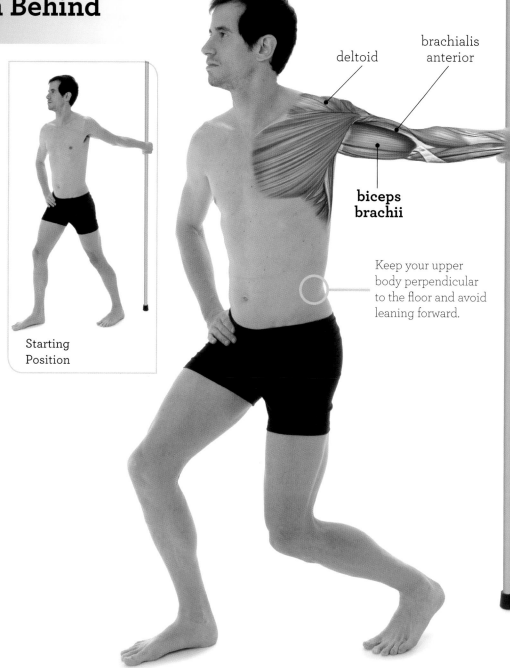

Starting Position

deltoid

brachialis anterior

biceps brachii

Keep your upper body perpendicular to the floor and avoid leaning forward.

CAUTION

Reduce the tension before you feel any pain in your elbow or shoulder, and remember that the discomfort in the stretch may come from muscle tension and not joint problems.

BENEFITS

Reduction of tension in the front part of your arm and shoulder, as well as maintenance of optimal joint motion in the shoulder and elbow.

INDICATION

For people who do heavy or mechanical work with their arms, or who keep their elbows bent for a long time, such as people who work at desks or computers. Also for strength and hypertrophy athletes.

Frontal Elbow Wall Support

triceps
brachii

deltoid
posterior

teres
major

latissimus
dorsi

Starting Position

Keep one foot ahead
of the other so you
can get closer to and
farther from the wall
with minimal deviation
from the starting
position.

START
Stand close to a wall
and facing it. Rest
the back part of your
elbow on the wall and
keep one leg slightly
to the rear. Remember
that your arm must
be relatively high
to achieve the right
support angle, and
your elbow needs to
be bent.

TECHNIQUE
Move closer to the wall
without moving your
feet, while sliding your
elbow upward along
it, until the greater
part of your arm is in
contact with it. Keep
your elbow bent as
much as possible
to create maximum
tension in your
triceps. The tension
in this exercise is very
perceptible.

LEVEL	REPS	DURATION
BEGINNER	2	20 sec.
INTERMEDIATE	3	25 sec.
ADVANCED	4	45 sec.

CAUTION
As in all stretches, especially
those involving the shoulder,
perform the movement
slowly and be alert to the
sensations in your joints.

BENEFITS
Reduction of tension in the
back part of the arm.

INDICATION
For people who do repetitive
movements straightening
the elbow or who have to
push heavy objects, such
as stretchers, wheelchairs,
carts, and the like.

Grip Behind Back

triceps brachii

deltoid posterior

Keep your hands together behind your back.

teres major

latissimus dorsi

START
Stand with your feet shoulder-width apart and raise one arm. You can bend the raised arm slightly and keep the other one relaxed at your side.

TECHNIQUE
Bend your elbows and try to clasp both hands behind your back. You probably will be able to hook together only your index, middle, and ring fingers. If you manage to do this, pull to increase the tension of the stretch in the back of the raised arm.

Starting Position

LEVEL	REPS	DURATION
BEGINNER	2	20 sec.
INTERMEDIATE	3	35 sec.
ADVANCED	4	45 sec.

CAUTION
As in the previous exercise, your shoulders will reach the limits of their range, so you will have to be particularly alert to the sensations they transmit.

BENEFITS
Reduction of tension in the back part of the arm.

INDICATION
For people who regularly push heavy items, such as carts, wheelbarrows, and the like.

Posterior Elbow Pull

For a better stretch, keep the elbow on the side being stretched bent to the maximum.

triceps brachii

deltoid posterior

teres major

latissimus dorsi

Starting Position

START
Stand and raise your arms. Bend one elbow completely, such that your hand is behind your head. The other hand grips the opposite elbow.

TECHNIQUE
Pull rearward on the elbow that is bent more. The stronger the pull, the more intense the stretch. You will easily feel the stretch, as in most triceps stretches. Hold the tension for a few seconds and return to the starting point.

LEVEL	REPS	DURATION
BEGINNER	2	20 sec.
INTERMEDIATE	3	35 sec.
ADVANCED	4	45 sec.

CAUTION
The shoulder on the side being stretched reaches one of its limits, so you will have to do the stretch slowly and be alert to the tiniest joint pain.

BENEFITS
Reduction of tension in the rear of the arm.

INDICATION
For people who regularly push heavy weights, such as carts, wheelbarrows, and the like.

Wrist Pull and Extension

START

Stand with both arms in front of you. One of your hands is palm up. It grasps the other, which is palm down.

TECHNIQUE

Pull downward on the first hand such that your wrist and elbow are totally extended, and slightly turn your wrist. You will feel the tension in your forearm, probably in the upper part, where the epitrochlear muscles are a bit thicker. Hold the tension for a few seconds and return to the starting position before repeating the exercise.

Starting Position

brachialis

epitrochlear muscles

The hand of the arm being stretched is in supination.

LEVEL	REPS	DURATION
BEGINNER	2	15 sec.
INTERMEDIATE	3	20 sec.
ADVANCED	4	35 sec.

CAUTION

Be careful with this stretch if you feel discomfort in the wrist area, for example, due to carpal tunnel syndrome.

BENEFITS

Reduction of tension in the front part of the forearm.

INDICATION

For people who do repetitive movements with their forearms or for people who do manual work that requires strength, such as mechanics, farmers, and massage therapists. Also for people who have epitrochleitis or golfer's elbow.

Bilateral with Inverted Support

Starting Position

brachialis

Your fingers must point rearward, toward your knees.

epitrochlear muscles

START
Get down on all fours on a mat, with your hands fairly close to your knees. Your palms must contact the mat with your fingers pointing toward your knees and your wrists slightly rotated.

TECHNIQUE
Slowly shift your weight rearward without changing any of the support points. Move rearward until you are sitting on your calves and your wrists are at maximum extension. Hold the point of maximum stretch for a few seconds.

LEVEL	REPS	DURATION
BEGINNER	2	15 sec.
INTERMEDIATE	3	20 sec.
ADVANCED	4	35 sec.

CAUTION	BENEFITS	INDICATION
As you move rearward, if you feel that the tension is excessive or you feel pain in your wrists, stop the movement or start over with your hands closer to your knees, which will reduce the wrist extension.	Reduction of tension in the front part of your forearms.	For people who do manual work requiring strength or repetitive movements and for people who suffer from epitrochleitis or golfer's elbow.

Bent Wrist Pull

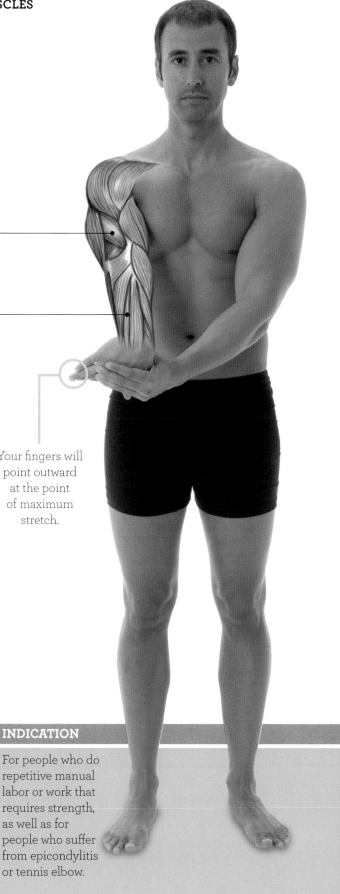

brachialis ———

epicondylar muscles ———

START
Stand with your arms to the front. One hand will be palm down and the other hand will hold it. The holding hand will act like tongs, with your thumb in the palm of the other hand and the rest of your fingers on the back.

TECHNIQUE
Pull the first hand downward and outward, such that the wrist bends and rotates outward. You will feel tension in the upper part of your forearm, which indicates that the movement is correct and that the stretch of the epicondylar muscles is being done properly.

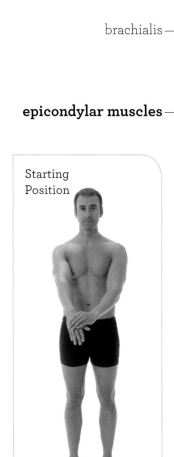

Starting Position

Your fingers will point outward at the point of maximum stretch.

LEVEL	REPS	DURATION
BEGINNER	2	15 sec.
INTERMEDIATE	3	20 sec.
ADVANCED	4	35 sec.

CAUTION
This stretch involves no risk beyond that of forcing the wrist joint, in which case you will feel pain before reaching an excessive bend.

BENEFITS
Reduction of tension in the rear of the forearm.

INDICATION
For people who do repetitive manual labor or work that requires strength, as well as for people who suffer from epicondylitis or tennis elbow.

Bilateral with Back Hand Support

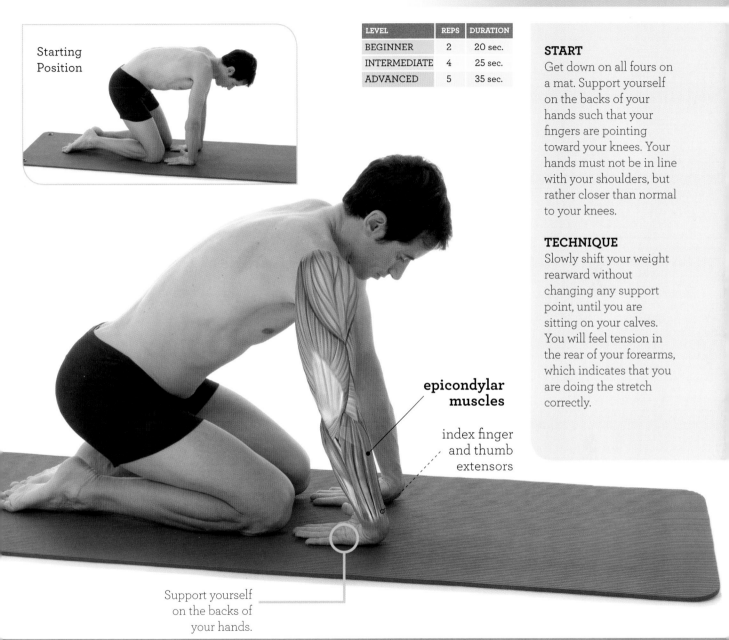

Starting Position

Support yourself on the backs of your hands.

epicondylar muscles

index finger and thumb extensors

LEVEL	REPS	DURATION
BEGINNER	2	20 sec.
INTERMEDIATE	4	25 sec.
ADVANCED	5	35 sec.

START
Get down on all fours on a mat. Support yourself on the backs of your hands such that your fingers are pointing toward your knees. Your hands must not be in line with your shoulders, but rather closer than normal to your knees.

TECHNIQUE
Slowly shift your weight rearward without changing any support point, until you are sitting on your calves. You will feel tension in the rear of your forearms, which indicates that you are doing the stretch correctly.

CAUTION
You probably will feel discomfort before you reach your calves. If so, you can stop the movement before reaching the point of pain or return to the starting position, moving your hands closer to your knees or even placing them next to your knees.

BENEFITS
Reduction of tension in the rear part of your forearm.

INDICATION
For people who do manual labor that requires repetitive movements or strength, as well as for people who suffer from epicondylitis or tennis elbow.

HAND AND WRIST STRETCHES

There are a great many muscles that are used to bend and straighten the wrist and hand. Here we will focus on the ones that are more directly involved with the hands and the fingers.

The following are the muscles that play a more significant role in bending the hand and the fingers:

Dorsal interossei muscles: These originate at the sides of the adjacent metacarpals, and they insert at the bases of the phalanges. Their function is the abduction and flexion of the index, middle, and ring fingers at the metacarpophalangeal joints.

Palmar interossei muscles: These muscles originate at the palmar surfaces of the metacarpals, and they insert at the bases of the phalanges. Their main functions are adduction and flexion of the index, ring, and little fingers.

Deep common flexor of the fingers: This muscle arises at the medial and proximal part of the ulna, and it inserts at the anterior surfaces of all the fingers, except for the thumb. It produces the flexion of the wrist, hand, and fingers, except for the thumb.

Superficial common flexor of the fingers: This muscle arises at the epitrochlea of the humerus and the anterior part of the ulna and the radius, and it inserts at the anterior surfaces of all the fingers, except for the thumb. It shares functions with the deep common flexor of the fingers.

Short flexor of the thumb: This originates at the flexor retinaculum flexorum manus and the trapezium bone, and inserts at the proximal phalange of the thumb. Its main function is flexion of the thumb.

Long thumb flexor: This originates at the anterior surface of the radius, and inserts at the base of the distal phalange of the thumb. Its function is flexion of the thumb and the wrist.

Lumbricals of the hand: These muscles originate at the distal tendons of the deep common flexor of the fingers, and insert at the distal tendons of the extensor of the fingers. Their function is flexion of all the fingers, except for the thumb, in the metacarpophalangeal joints and the extension of these in the interphalangeal joints.

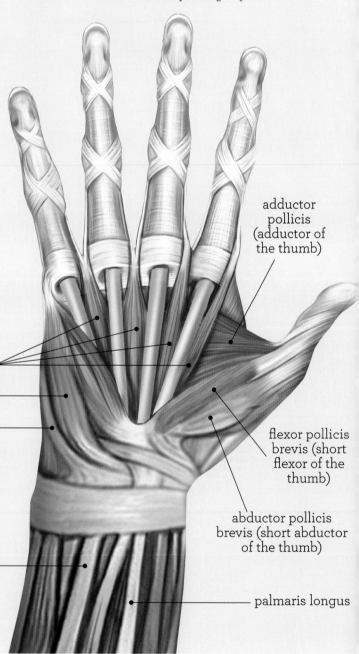

adductor pollicis (adductor of the thumb)

musculi lumbricales manus (lumbrical muscles of the hand)

flexor digiti minimi (flexor of the little finger)

abductor digiti minimi (abductor of the little finger)

flexor pollicis brevis (short flexor of the thumb)

abductor pollicis brevis (short abductor of the thumb)

flexor digitorum superficialis communis (superficial common flexor of the fingers)

palmaris longus

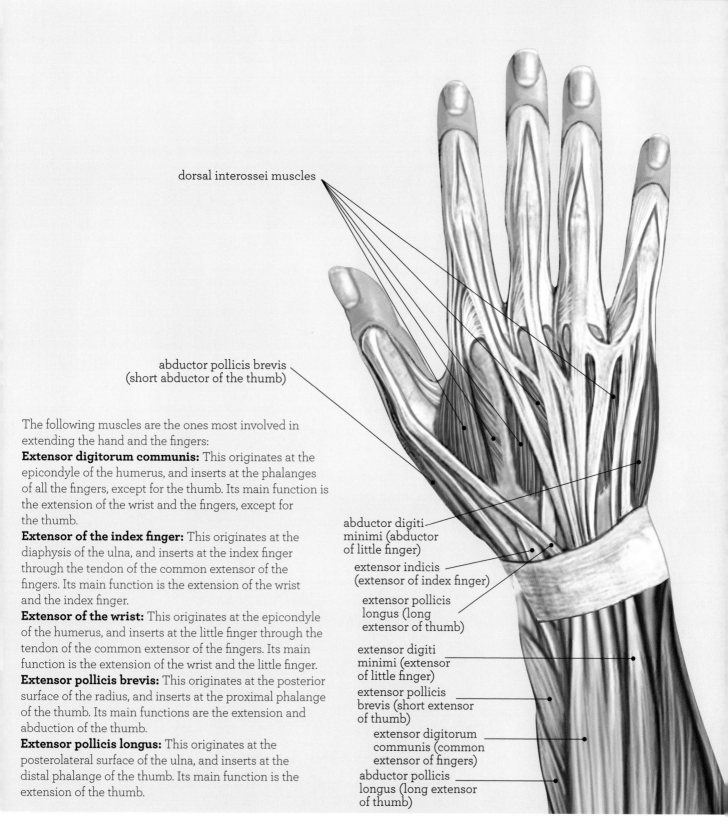

dorsal interossei muscles

abductor pollicis brevis
(short abductor of the thumb)

abductor digiti
minimi (abductor
of little finger)

extensor indicis
(extensor of index finger)

extensor pollicis
longus (long
extensor of thumb)

extensor digiti
minimi (extensor
of little finger)

extensor pollicis
brevis (short extensor
of thumb)

extensor digitorum
communis (common
extensor of fingers)

abductor pollicis
longus (long extensor
of thumb)

The following muscles are the ones most involved in extending the hand and the fingers:

Extensor digitorum communis: This originates at the epicondyle of the humerus, and inserts at the phalanges of all the fingers, except for the thumb. Its main function is the extension of the wrist and the fingers, except for the thumb.

Extensor of the index finger: This originates at the diaphysis of the ulna, and inserts at the index finger through the tendon of the common extensor of the fingers. Its main function is the extension of the wrist and the index finger.

Extensor of the wrist: This originates at the epicondyle of the humerus, and inserts at the little finger through the tendon of the common extensor of the fingers. Its main function is the extension of the wrist and the little finger.

Extensor pollicis brevis: This originates at the posterior surface of the radius, and inserts at the proximal phalange of the thumb. Its main functions are the extension and abduction of the thumb.

Extensor pollicis longus: This originates at the posterolateral surface of the ulna, and inserts at the distal phalange of the thumb. Its main function is the extension of the thumb.

Wrist and Finger Flex

START
Place your hands in front of you with the palms facing upward. One hand covers the back of the other in such a way that the fingers of each hand point toward the opposite shoulder.

TECHNIQUE
Press one hand against the other, so that your wrist and metacarpophalangeal joints are bent. As you approach the end of the motion, you will feel tension in the back of your hand, which will indicate that the stretch is taking place.

Starting Position

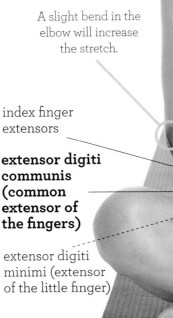

A slight bend in the elbow will increase the stretch.

index finger extensors

extensor digiti communis (common extensor of the fingers)

extensor digiti minimi (extensor of the little finger)

LEVEL	REPS	DURATION
BEGINNER	2	20 sec.
INTERMEDIATE	3	25 sec.
ADVANCED	4	30 sec.

CAUTION
Even though the wrist joint is not as unstable as the shoulder, it is a small joint composed of tiny bones, so you should not subject it to excessive tension.

BENEFITS
Reduction of tension in the back of the hand and the posterior area of the forearm.

INDICATION
For people who use their fingers and hands in repetitive activities, such as typing on a computer keyboard, painting, sewing, and the like.

Assisted Wrist Extension

Starting Position

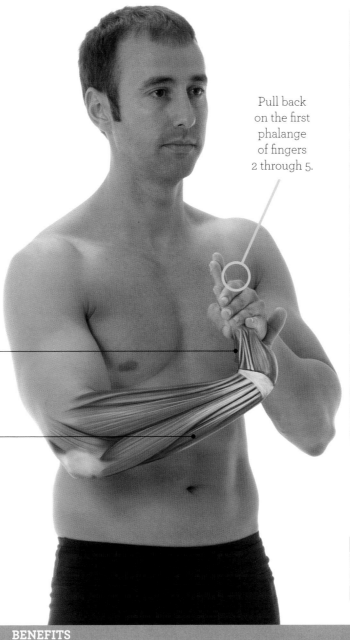

Pull back on the first phalange of fingers 2 through 5.

interossei and lumbricales of the hand

flexores digitorum communes (common flexors of the fingers

START
Place both hands in front of you. Hold one hand with the other such that the thumb is over the first phalange of the fingers on the opposite hand, except for the thumb, and the rest of the fingers are beneath the phalanges.

TECHNIQUE
With the hand doing the holding, pull on the other hand to produce an extension of the wrist and metacarpophalangeal joints. The tension at the base of the fingers and the anterior part of the forearm will indicate that you are doing the stretch.

LEVEL	REPS	DURATION
BEGINNER	2	20 sec.
INTERMEDIATE	3	25 sec.
ADVANCED	4	30 sec.

CAUTION
As in the previous case, do not force the extension of the wrist beyond the discomfort that goes along with any stretch.

BENEFITS
Reduction of tension in the anterior part of the forearm, the palm of the hand, and the fingers.

INDICATION
For people who repeatedly use their hands and fingers in office or manual work, as well as those who hold or manipulate heavy weights with their hands.

Finger Extension

interosei and lumbricals of the hand

flexor digitorum communis

Keep your elbows extended during the stretch.

START

Stand or sit, interlace your fingers, and without releasing them, point the palms of your hands outward as you stretch your arms. At this point, you will begin to feel a certain tension that will increase as you continue the stretch.

TECHNIQUE

Keeping your fingers interlaced, raise your hands until they are just over your head with your palms pointing at the ceiling. If you feel enough tension at any point between the starting position and the final one, you can stop the movement and hold the stretch. The optimal point will vary from person to person.

Starting Position

LEVEL	REPS	DURATION
BEGINNER	2	15 sec.
INTERMEDIATE	3	20 sec.
ADVANCED	3	30 sec.

CAUTION

Make sure that the extension of the fingers does not produce any type of lateral traction or tension in them, because it might involve a risk to the integrity of the joints.

BENEFITS

Reduction of tension in the fingers, the palms of the hands, and the anterior part of the forearm, avoidance of strain, and maintenance of optimal joint mobility.

INDICATION

For people who use their fingers for long periods of time, whether keyboarding at a computer, using a mouse, writing or drawing, and even playing a piano or keyboard. Also for those who do work that requires strength in the hands and fingers, or the use of tools, such as hammers, screwdrivers, wrenches, and so forth.

Thumb Bend

abductores pollicis (thumb abductors)

extensores pollicis (thumb extensors)

Clench your fist with your thumb inside.

Starting Position

START
Place one hand in front of you and wrap your fingers around your thumb, as if you wanted to hold it in your palm.

TECHNIQUE
Clench your fist tightly to force the flexion of the thumb in its metacarpophalangeal joint while tipping your knuckles toward the floor using wrist adduction.

LEVEL	REPS	DURATION
BEGINNER	2	20 sec.
INTERMEDIATE	3	25 sec.
ADVANCED	4	30 sec.

CAUTION
This exercise involves no risks, because it does not allow enough forced movement to damage joints.

BENEFITS
Reduction of tension in the thumb region.

INDICATION
For people who do repetitive tasks or ones that involve the fingers to a large degree, such as keyboarding and painting or drawing, and for people who hold or manipulate heavy items with their hands.

Rhombus Position

START

Place your hands in front of you such that your palms face your body and their backs point straight ahead. Then bring your hands together and join your index finger and thumb on one hand with their counterparts on the other such that there is a space in the middle that resembles a rhombus.

TECHNIQUE

Press one hand against the other such that the rhombus-shaped space becomes longer and narrower. The tension in the area between the index finger and thumbs and the base of the latter will be a good indicator that the stretch is happening.

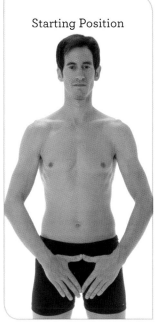

Starting Position

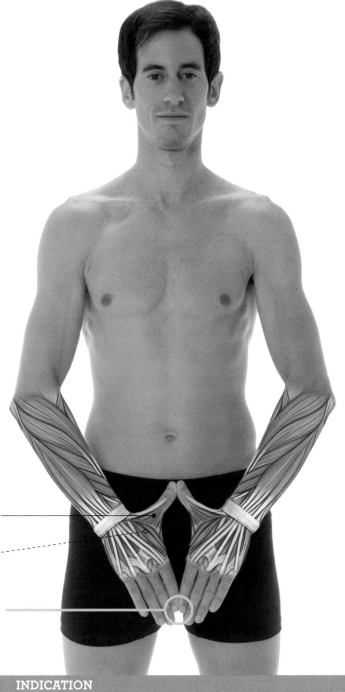

flexores pollicis (thumb flexors)

adductor pollicis (thumb adductor)

Keep the index fingers and thumbs of both hands in contact with one another.

LEVEL	REPS	DURATION
BEGINNER	2	20 sec.
INTERMEDIATE	3	25 sec.
ADVANCED	4	30 sec.

CAUTION

Make sure that the contact between both hands is firm before starting to exert pressure.

BENEFITS

Reduction of tension in the thumb region.

INDICATION

For people who do manual work or habitually use their hands and fingers, such as painters, illustrators, and office employees, and individuals who travel a lot by bicycle or motorcycle. Also for people who handle or hold heavy weights in their hands.

Thumb Pull

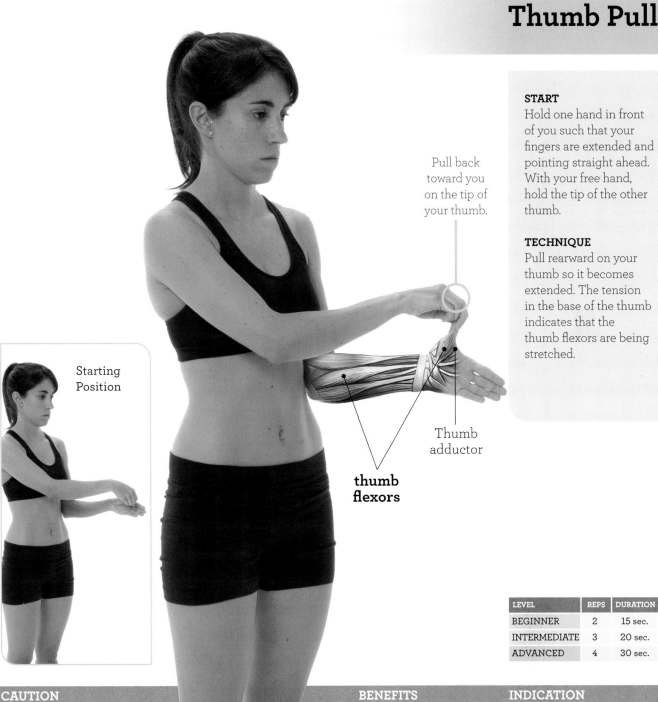

Pull back toward you on the tip of your thumb.

Starting Position

Thumb adductor

thumb flexors

START
Hold one hand in front of you such that your fingers are extended and pointing straight ahead. With your free hand, hold the tip of the other thumb.

TECHNIQUE
Pull rearward on your thumb so it becomes extended. The tension in the base of the thumb indicates that the thumb flexors are being stretched.

LEVEL	REPS	DURATION
BEGINNER	2	15 sec.
INTERMEDIATE	3	20 sec.
ADVANCED	4	30 sec.

CAUTION
Do the stretch by applying low to moderate tension, because you are moving very small joints.

BENEFITS
Reduction of tension in the thumb area.

INDICATION
For people who do repetitive tasks with their hands and fingers, travel by motorcycle or bicycle for a long time, or who regularly hold or use heavy objects.

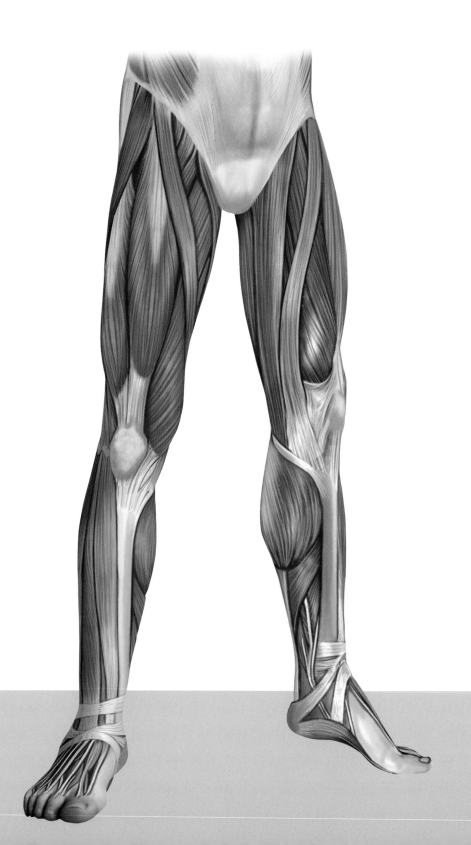

STRETCHES

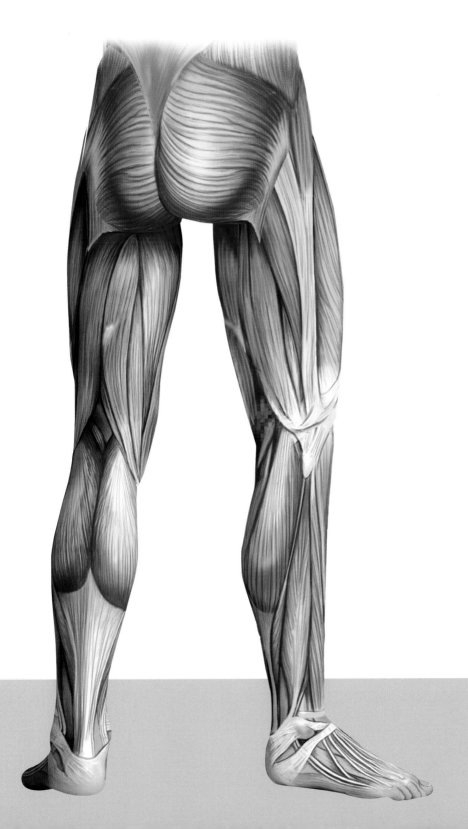

HIP STRETCHES

ADDUCTORS
These are the major, medial, and minor adductors. All three have a common origin in the pubis, and they insert along the diaphysis of the femur. Their main function is hip adduction, which amounts to moving the lower extremity toward the center line of the body, thus making it possible to bring the legs together.

TENSOR FASCIAE LATAE
This muscle originates at the iliac crest, the fascia lata, and the iliac spine. It inserts at the proximal epiphysis of the tibia. Its function is hip abduction, mainly starting from the position of flexion, and therefore it moves the lower extremity away from the center line of the body.

ILIAC
This muscle originates at the internal area of the ilium, and it inserts at the lesser trochanter of the femur. Its main functions are the flexion and lateral rotation of the hip.

PSOAS MAJOR
This muscle originates at the transverse apophyses of vertebrae T12 through L5 and at their intervertebral discs, and they insert at the trochanter minor of the femur. Its major function is hip flexion.

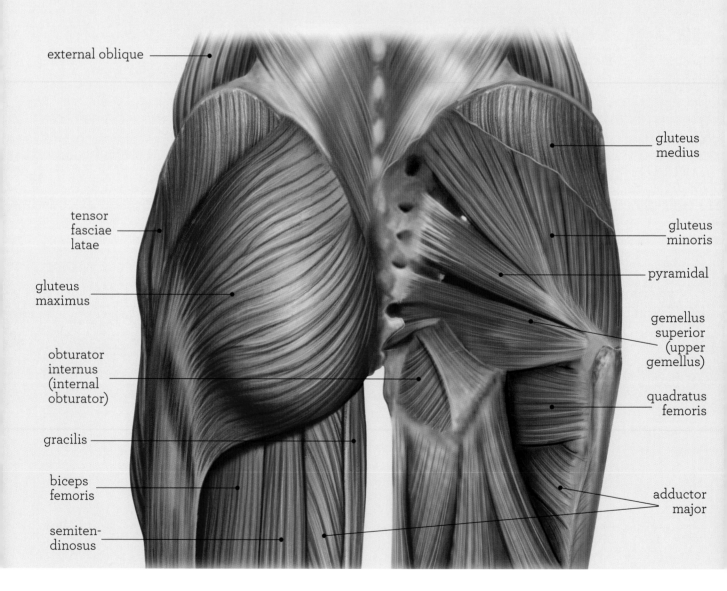

external oblique

gluteus medius

tensor fasciae latae

gluteus minoris

pyramidal

gluteus maximus

gemellus superior (upper gemellus)

obturator internus (internal obturator)

quadratus femoris

gracilis

biceps femoris

adductor major

semiten-dinosus

GLUTEALS

These are the gluteus maximus, medius, and minimus.

Gluteus maximus: This muscle originates in the ilium, sacrum, and coccyx bones, and it inserts at the proximal third of the femur. Its main function is hip extension.

Gluteus medius: This muscle originates at the posterior part of the iliac crest, and it inserts at the trochanter major of the femur. Its main function is hip abduction, which it shares with the gluteus minimus and, to a lesser degree, with the tensor fasciae latae.

Gluteus minimus: This muscle originates at the external ilium, and it inserts at the trochanter major of the femur. Its main function is hip abduction.

PYRAMIDAL

This muscle originates at the sacrum bone, and inserts at the trochanter major of the femur. Its main function is the lateral rotation of the hip.

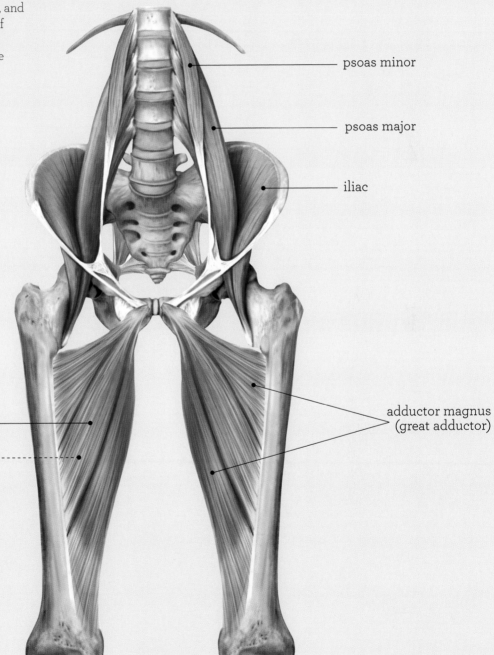

psoas minor

psoas major

iliac

adductor medius
(medium adductor)

adductor minimus
(small adductor)

adductor magnus
(great adductor)

Standing Leg Extension

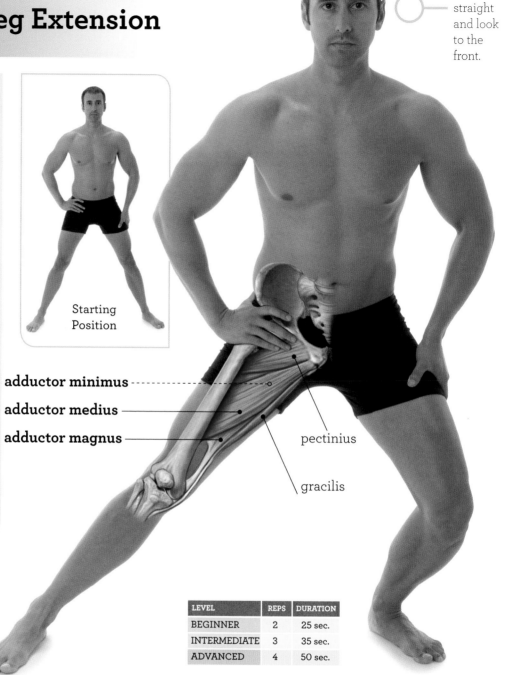

Keep your back straight and look to the front.

START
Stand with your feet apart, the hand on the side to be stretched resting on your waist and the other hand on the top of your thigh, in order to contribute to comfort and stability during the stretch.

TECHNIQUE
Bend the knee of the leg that is not being stretched, so that your center of gravity lowers and moves toward that side. As you do the movement, you will feel the tension in the inside of the thigh being stretched, which indicates that tension is being created in your adductors and, therefore, that you are doing the movement correctly.

Starting Position

adductor minimus
adductor medius
adductor magnus

pectinius

gracilis

LEVEL	REPS	DURATION
BEGINNER	2	25 sec.
INTERMEDIATE	3	35 sec.
ADVANCED	4	50 sec.

CAUTION
Perform the movement gradually and progressively, because these muscles are not excessively strong, which makes them more susceptible to injury from sudden movements.

BENEFITS
Broadening the range of movement and reduction of tension in the muscles on the inside thigh muscles.

INDICATION
For people who do physical activity, particularly involving the lower body, or who suffer from specific disorders of the hips and pelvis, such as athletic pubalgia.

Leg Extension on All Fours

Starting Position

LEVEL	REPS	DURATION
BEGINNER	2	25 sec.
INTERMEDIATE	3	35 sec.
ADVANCED	4	50 sec.

START
Get down on all fours on a mat and stretch one leg out to the side so that your foot is supported on its inside surface. The three other support points are your two hands and your knee on the other leg.

TECHNIQUE
Slide the supported foot on the mat in such a way that it moves away from the other foot and approaches maximum extension. You will soon feel the tension in the inside area of your thigh, which indicates that the adductors are being stretched.

adductor minimus

adductor medius

adductor magnus

gracilis

pectinius

The foot of the leg being stretched slides along on its inner surface.

CAUTION	BENEFITS	INDICATION
Do the stretch slowly and progressively, keeping the three support points firm.	Broadening the range of motion and reduction of tension in the inside thigh muscles.	For people who do physical activity or suffer from specific hip and pelvis problems, such as athletic pubalgia.

Standing Leg Raise

START
Stand next to a raised object, such as a step, a box, or a small stool. Raise the leg closest to the item and rest the inside of your foot on it.

TECHNIQUE
Bend the knee of the support leg in such a way that your center of gravity lowers and the abduction of the raised leg increases. As with previous stretches, you will feel tension in the inside of the thigh.

Keep your back straight and look to the front.

Starting Position

adductor medius

adductor minimus

pectinius

adductor magnus

gracilis

LEVEL	REPS	DURATION
BEGINNER	2	25 sec.
INTERMEDIATE	3	30 sec.
ADVANCED	4	40 sec.

CAUTION
Be sure to start from a balanced position and to do the movement slowly so you don't lose your balance.

BENEFITS
Broadening of the range of movement and reduction of tension in the inside thigh muscles.

INDICATION
Like the other adductor stretches, for improving the condition of patients with hip and pelvis irregularities, especially athletes, such as soccer players, who tend toward shortening of the adductors or toward injury.

Bilateral in Sumo Position

Starting Position

Rest your elbows on your thighs, just above your knees.

pectinius

adductor medius

adductor minimus

adductor magnus

gracilis

START
Stand with your feet greater than shoulder-width apart. Bend your knees to 100 or 110°. Lean your upper body forward and rest your elbows on your thighs, just above your knees.

TECHNIQUE
Lower your chest and push your knees outward with your elbows, so that the knees separate farther. This movement will produce tension in your adductors, which you should maintain for a few seconds before finishing the stretch.

LEVEL	REPS	DURATION
BEGINNER	2	20 sec.
INTERMEDIATE	4	25 sec.
ADVANCED	5	35 sec.

CAUTION	BENEFITS	INDICATION
Do this stretch slowly, and start from the most stable position possible, because the stretch will move your center of gravity, and it may make you lose your balance.	Broadening of the range of movement and reduction of tension in the muscles on the inside of the thighs.	For people with hip disorders and who do athletic disciplines that require high performance from the lower body, especially explosive movements or sudden changes of direction.

Backward Movement on Knees and Forearms

START

Take a position on a mat, supporting yourself on your knees and forearms. Keep your elbows shoulder-width apart or slightly closer, and your knees considerably farther apart than your elbows.

TECHNIQUE

Move your body rearward without moving your support points, such that your elbows extend and your gluteals end up over your heels. You will feel tension in the inside of your thighs, which you should hold for a few seconds.

Starting Position

LEVEL	REPS	DURATION
BEGINNER	2	20 sec.
INTERMEDIATE	3	25 sec.
ADVANCED	4	35 sec.

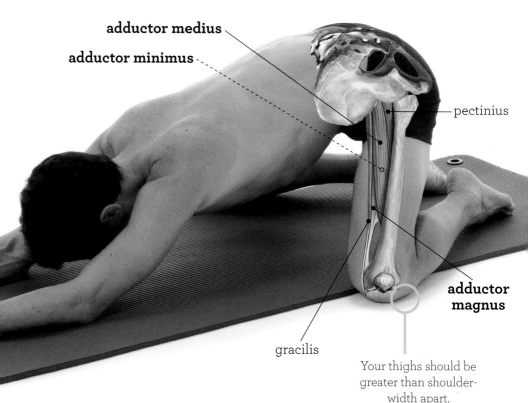

adductor medius

adductor minimus

pectinius

adductor magnus

gracilis

Your thighs should be greater than shoulder-width apart.

CAUTION	BENEFITS	INDICATION
If you feel pain in your hip joint, use one of the other suggested stretching exercises for the adductors.	Broadening the range of movement and reduction of tension in the muscles on the inside of the thigh.	For athletes in sports that place high demands on the lower body and require explosive efforts, such as changes of direction, sudden starts, sudden stops, jumps, especially tennis players, soccer players, and skaters.

Butterfly Position

Starting Position

LEVEL	REPS	DURATION
BEGINNER	2	20 sec.
INTERMEDIATE	3	25 sec.
ADVANCED	4	35 sec.

START
Sit on a mat with the soles of your feet together. Place your hands on your ankles and your elbows in the inside of your thighs.

TECHNIQUE
Press outward with your elbows so your knees move farther apart but the soles of your feet remain together. You will feel the tension from the stretch on the inside of your thighs. Hold the position for a few seconds before returning to the starting point.

adductor minimus

adductor medius

pectineus

Press on the inside of your thighs with your elbows.

adductor magnus

gracilis

CAUTION
Avoid sudden or rapid movements in doing this stretch. Remember that only slow movement will let you react to a painful sensation.

BENEFITS
Broadening the range of movement and reduction of tension in the muscles on the inside of your thighs.

INDICATION
For people who engage in sports that place high demands on the lower body or that involve a restricted range of movement in that part of the body.

Seated Legs in V-position

START
Sit on a mat with your legs straight and apart. Your hands should be between your legs at the start of the movement.

TECHNIQUE
Gradually separate your legs even more with the help of your hands pushing outward. Your upper body will need to lean forward while you push your legs apart.

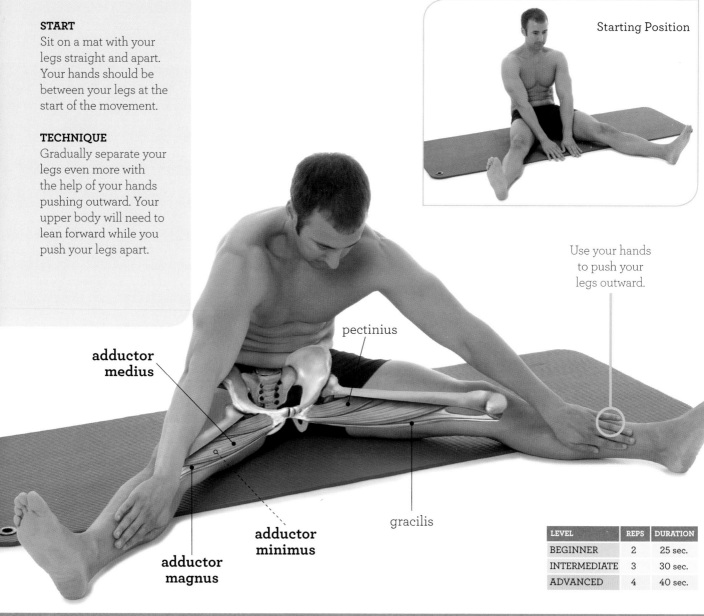

Starting Position

Use your hands to push your legs outward.

pectinius

adductor medius

gracilis

adductor minimus

adductor magnus

LEVEL	REPS	DURATION
BEGINNER	2	25 sec.
INTERMEDIATE	3	30 sec.
ADVANCED	4	40 sec.

CAUTION

Do this stretch slowly and be alert to your back as well as to your adductors, because the bend in your spine could produce some small localized discomfort if you have back pain or preexisting irregularities.

BENEFITS

Broadening of range of movement and prevention of internal injuries to the internal thigh muscles.

INDICATION

For athletes with tight adductors or who are looking for a greater range of movement due to the requirements of their sports.

Supine with Legs in V-position

Starting Position

START
Lie down on a mat, straighten your legs, and extend them upward and slightly apart. Place your hands on the inside of your thighs and rest your head on the floor.

TECHNIQUE
Slowly and gradually spread your legs apart. Use your hands to increase the tension and optimize the effects of the stretch.

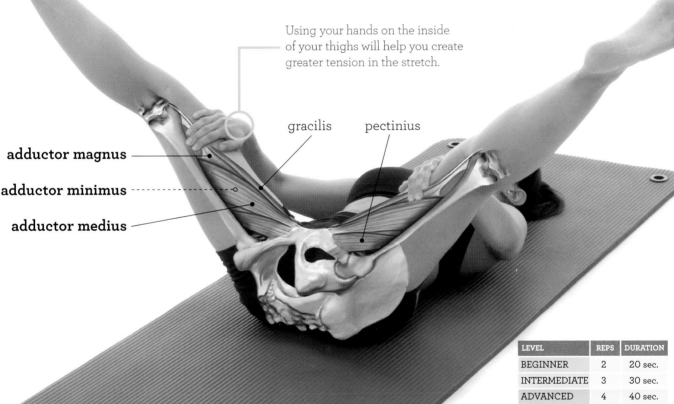

Using your hands on the inside of your thighs will help you create greater tension in the stretch.

gracilis

pectinius

adductor magnus

adductor minimus

adductor medius

LEVEL	REPS	DURATION
BEGINNER	2	20 sec.
INTERMEDIATE	3	30 sec.
ADVANCED	4	40 sec.

CAUTION
Keep your head and neck relaxed and in contact with the mat, and stop the stretch as soon as you begin to feel pain. Remember that the sensation of tension and discomfort, without reaching the point of pain, is desirable in flexibility work.

BENEFITS
Broadening of range of movement and reduction of tension in the muscles on the inside of the thighs.

INDICATION
For people who have tight adductors or little general flexibility in their lower bodies and for athletes who require greater ranges of movement.

Lateral Torso Flex with Crossed Leg

START
Take a standing position and cross the leg to be stretched behind the other. Rest the hand corresponding to the crossed leg on your hip and let the other arm hang loosely by your side.

TECHNIQUE
Slide your forward foot along the floor so it crosses the rear leg even more. At the same time, bend your upper body in the opposite direction, letting your free arm hang down.

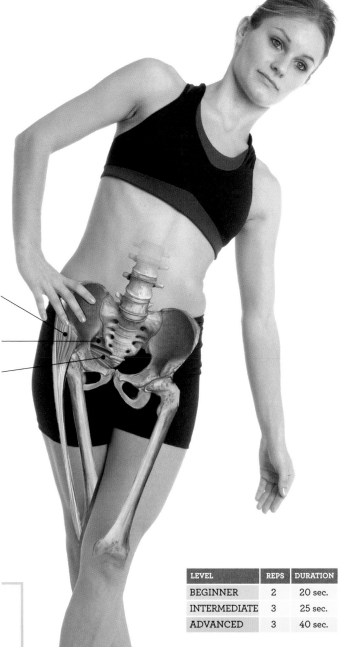

tensor fasciae latae

gluteus medius

gluteus minimus

Starting Position

Slide the forward foot away from the rear one while keeping your legs crossed.

LEVEL	REPS	DURATION
BEGINNER	2	20 sec.
INTERMEDIATE	3	25 sec.
ADVANCED	3	40 sec.

CAUTION
You may not feel the tension, or at least not to the same degree as in the adductor stretches. This does not necessarily mean that you are not doing the exercise properly or that you have to prolong the movement excessively.

BENEFITS
Reduction of tension in the muscles of the outer thigh and increase in range of movement.

INDICATION
For people who do physical activity that places special demands on the lower body or who have specific hip or knee disorders, such as iliotibial band syndrome or runner's knee.

Unilateral Standing with Support

START
Stand with your side facing a support. Keep your feet together and one hand on the chosen support point to keep your balance during the execution of this exercise.

TECHNIQUE
Bend your upper body to the side and toward the support point as you thrust your hip toward the opposite side. Keeping your feet in place, slightly bend the knee of the inside leg to increase the projection of the hip.

tensor fasciae latae

gluteus medius

gluteus minimus

Bend the inside knee to increase the stretch.

Starting Position

LEVEL	REPS	DURATION
BEGINNER	2	20 sec.
INTERMEDIATE	3	25 sec.
ADVANCED	3	40 sec.

CAUTION
You should not increase the movement excessively if you don't feel the tension, because it does not manifest itself as clearly as in stretches for other muscle groups.

BENEFITS
Broadening of range of movement and reduction of tension in the muscles on the outside of the thigh and hip.

INDICATION
For athletes whose disciplines involve the lower body and people with specific hip and knee disorders, such as iliotibial band syndrome or runner's knee.

Unilateral with Step

START

Stand in front of a stool, chair, or other object with a raised surface. Rest one foot on the stool and keep the other one on the floor, aligned with your hip and bearing your weight. You can place your hands on your hips or let your arms hang loose at your sides.

TECHNIQUE

Move your body forward without moving your feet from their support points, while keeping your upper body perpendicular to the floor. As you move forward, your center of gravity will lower, and the hip corresponding to your rear foot will increase its extension, thereby producing the stretch.

Move forward and lower your center of gravity to increase the extension of your hip.

psoas major

iliac

sartorius

gracilis

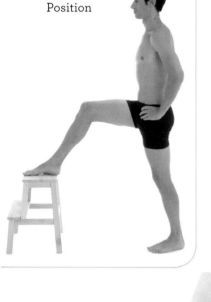

Starting Position

LEVEL	REPS	DURATION
BEGINNER	2	30 sec.
INTERMEDIATE	3	35 sec.
ADVANCED	4	45 sec.

CAUTION

Make sure the support point of the forward foot is stable and remains firmly planted during the stretch.

BENEFITS

Broadening of the range of movement and reduction of tension in the muscles in front of the hip.

INDICATION

For people who take part in physical activity or who suffer from specific hip disorders, hyperlordosis, or nonacute lower back problems.

Knight's Position

LEVEL	REPS	DURATION
BEGINNER	2	30 sec.
INTERMEDIATE	3	35 sec.
ADVANCED	4	45 sec.

START

Take a position on a mat, resting on one knee and one foot. The position is similar to kneeling down like a medieval knight while being granted knighthood—hence, the name of this exercise. The forward leg maintains a bend in the hip and knee, both at 90°. At the start, the rear leg keeps the thigh aligned with your upper body and the knee bent at 90°.

TECHNIQUE

Move forward without moving the anchor points such that you accentuate the extension of the hip corresponding to the rear leg. Your upper body must remain perpendicular to the floor during the exercise. Stop the movement when you feel the tension in your extended hip.

Starting Position

Keep your upper body perpendicular to the floor during the entire stretch.

psoas major

iliac

sartorius

gracilis

CAUTION

Make sure to begin with a stable position that will allow you to keep your balance during the entire exercise. Use a padded mat to avoid putting too much pressure on the support knee.

BENEFITS

Broadening of the range of movement and reduction of tension in the muscles in the front of the hip.

INDICATION

For people who do physical activity and for people with specific hip disorders, hyperlordosis, or nonacute lower back problems.

Knee Bend and Opposite Leg Extension

Starting Position

START

Lie down on your back on a bench, cot, or the like, with your legs pulled up and both hands on one knee. Keep your head on the bench and your neck relaxed.

TECHNIQUE

Lower the free leg to the maximum extension your hip will allow as you hold onto the other, which will remain in total hip and knee flexion. The knee of the lower leg must be almost totally extended, which will increase the stretch through the effect of gravity.

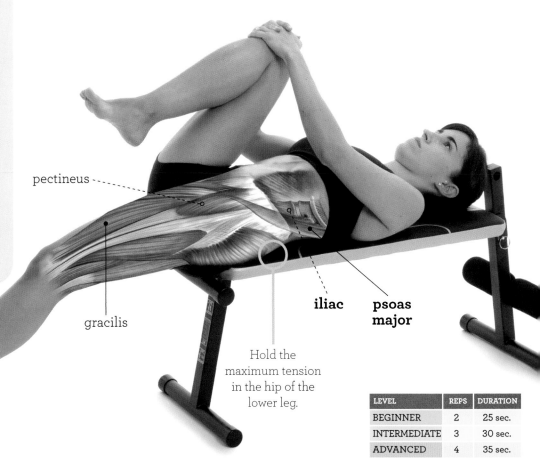

pectineus

gracilis

iliac

psoas major

Hold the maximum tension in the hip of the lower leg.

LEVEL	REPS	DURATION
BEGINNER	2	25 sec.
INTERMEDIATE	3	30 sec.
ADVANCED	4	35 sec.

CAUTION

Use a stable position and lower your leg in a gradual, controlled way to avoid sudden jerks and their possible consequences.

BENEFITS

Broadening of the range of movement and reduction of tension in the muscles in the front of the hip.

INDICATION

For people with specific hip disorders, hyperlordosis, and nonacute lower back problems, and athletes in general, especially those who experience tightness in the flexor muscles of the hip.

Knee Bend with Opposite Heel Rearward

Starting Position

START

Lie on your back on a workout bench, cot, or other raised surface that is not too low. Keep your legs drawn up and hold one knee with both hands. Rest your head on the bench and keep your neck relaxed.

TECHNIQUE

Extend one hip while you bend the knee of the corresponding leg. The sole of the lower foot must point rearward in such a way that the leg is drawn downward and rearward. The upper leg remains bent at the hip and the knee, and it is supported by both hands, which you use to pull it toward your chest.

gracilis

sartorius

iliac

psoas major

Keep your lumbar region close to the bench to avoid exaggerating the natural curvature of your spine.

LEVEL	REPS	DURATION
BEGINNER	2	25 sec.
INTERMEDIATE	3	30 sec.
ADVANCED	4	35 sec.

CAUTION

Avoid overextending your lumbar vertebrae as you do this exercise. Remain as close as possible to the bench.

BENEFITS

Broadening of the range of movement and reduction of tension in the muscles in the front of the hip.

INDICATION

For people with tight hip flexor muscles or who suffer from nonacute lower back problems or other hip disorders.

Unilateral with Crossed Leg

START

Sit on a mat with one leg straight and the other one crossed over it with knee bent. Hold this knee with both hands and keep your upper body perpendicular to the floor.

TECHNIQUE

Pull the bent knee toward the opposite side while keeping the supported foot anchored in place. You will feel the tension in the outside of your thigh and your gluteals, which indicates that the stretch is happening.

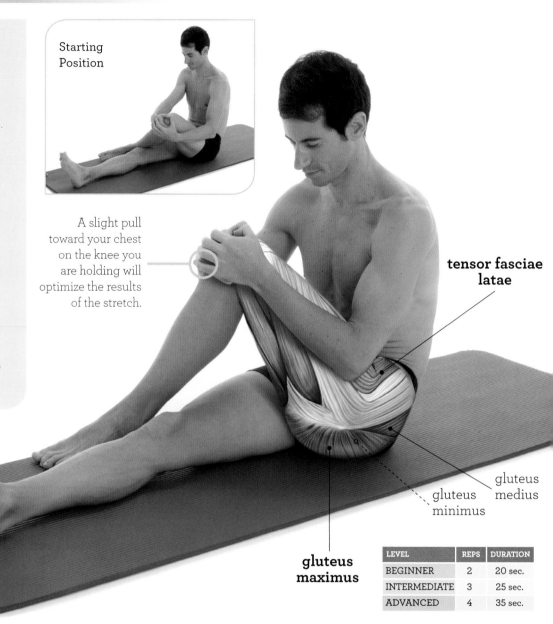

Starting Position

A slight pull toward your chest on the knee you are holding will optimize the results of the stretch.

tensor fasciae latae

gluteus medius

gluteus minimus

gluteus maximus

LEVEL	REPS	DURATION
BEGINNER	2	20 sec.
INTERMEDIATE	3	25 sec.
ADVANCED	4	35 sec.

CAUTION

Keep the supported foot anchored in one spot. Otherwise, you will not produce the desired stretch.

BENEFITS

Reduction of tension in the muscles in the rear of the hip and broadening of your range of movement.

INDICATION

For people who experience tightness in the extensor and abductor muscles of the hip, as well as for people who experience disorders of this joint.

Supine Knee and Hip Flex

Starting Position

START
Lie down on a mat, raise one leg, and bend your knee. Using the opposite hand, hold the outside of the raised knee. The other leg must be straight and aligned with your upper body.

TECHNIQUE
Pull your knee toward the inside such that it crosses over the other leg. You will feel tension in the side of your gluteal, which indicates that the stretch is happening properly. Hold the position for the appropriate time for your level before returning to the starting point.

Push the outside of the raised knee with the hand of the opposite side.

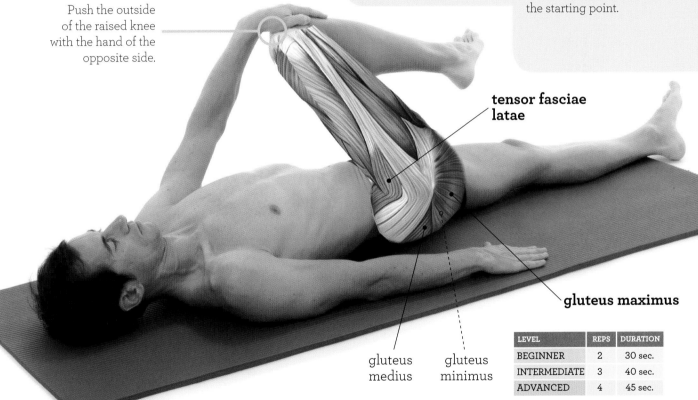

tensor fasciae latae

gluteus maximus

gluteus medius

gluteus minimus

LEVEL	REPS	DURATION
BEGINNER	2	30 sec.
INTERMEDIATE	3	40 sec.
ADVANCED	4	45 sec.

CAUTION
Keep your head in contact with the mat to avoid neck tension.

BENEFITS
Increase in range of movement and relaxation of the muscles in the rear of the hip.

INDICATION
For people with hip disorders and athletes in general.

Cross on Knee

START
Stand in front of a stool, chair, work surface, or any other support surface. Bend one knee and rest the outside of the ankle on the opposite thigh, just above the knee. Hold onto the selected support with both hands.

TECHNIQUE
Bend the knee of your support leg while you keep the other foot anchored in place and your hands holding the fixed point. As your center of gravity lowers, you will feel the tension that is produced in your gluteal due to the stretching exercise.

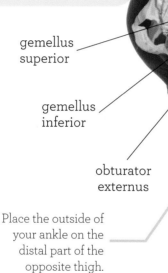

pyramidal

gemellus superior

gemellus inferior

obturator externus

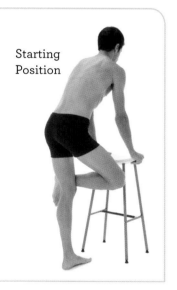

Starting Position

Place the outside of your ankle on the distal part of the opposite thigh.

LEVEL	REPS	DURATION
BEGINNER	2	20 sec.
INTERMEDIATE	3	25 sec.
ADVANCED	3	30 sec.

CAUTION
Make sure you hold on tight with both hands, because the stretch produces uncertain balance.

BENEFITS
Increase in range of movement and relaxation of the muscles in the rear of the hip.

INDICATION
For people who engage in physical activity and those who suffer from specific hip disorders or functional limitation due to muscle tightness in this joint.

Cross on Fixed Point

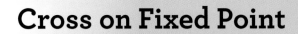

Starting Position

Push the raised knee outward and downward in order to optimize the stretch.

pyramidal

gemellus superior

gemellus inferior

obturator externus

START
Take a position in front of a stool or other raised support point. Place one foot on it so that it acts as a support and a solid stop. The other leg, which is straight, is the one that supports your weight. Place one hand on your knee and hold your ankle with the other.

TECHNIQUE
Slowly bend the knee of the leg that's holding you up while you push the other leg toward the outside such that the sole of your foot loses contact with the support and the outside of your ankle takes on the support function. As you do the movement, you will feel tension in the back of your hip.

LEVEL	REPS	DURATION
BEGINNER	2	20 sec.
INTERMEDIATE	3	25 sec.
ADVANCED	3	30 sec.

CAUTION
Be sure to start from a stable and balanced position, because the exercise will reduce your stability.

BENEFITS
Increase in range of movement and relaxation of the muscles in the back of the hip.

INDICATION
For athletes and for people who suffer from certain hip disorders or from muscle tightness in this joint.

Seated Cross

Starting Position

LEVEL	REPS	DURATION
BEGINNER	2	20 sec.
INTERMEDIATE	3	25 sec.
ADVANCED	3	30 sec.

START

Sit on a mat with one leg straight and the other bent at the knee so that your foot is next to your thigh. Place one hand on the bent knee and hold your foot with the other hand so that the sole of your foot and the palm of your hand are in contact with one another.

TECHNIQUE

Pull up on your foot while pushing the bent knee toward the floor. You will feel the tension in the rear of your hip. Continue the movement without moving from discomfort to pain.

Keep your back straight during the entire stretch.

gemellus superior

gemellus inferior

obturator externus

pyramidal

CAUTION

Do not force the stretch too much, because the pyramidal muscle is not very powerful, so it is more susceptible to injury if it is subjected to excessive tension.

BENEFITS

Increase in the range of movement and relaxation of the muscles in the rear of the hip.

INDICATION

For athletes and for people who suffer from specific hip disorders or functional limitatation due to muscle tightness in this joint.

Face-down Cross

Starting Position

START

Take a position face down on a mat, as if you were going to do a push-up. Support yourself with both hands and keep your elbows straight. Stretch out one leg and bend the other one at the hip and knee so that it crosses beneath you.

TECHNIQUE

Lean your upper body forward to bring your chest as close as possible to the floor, while sliding your hands forward. This will produce maximum extension in the hip of the crossed leg, and, ultimately, it will stretch the pyramidal muscle.

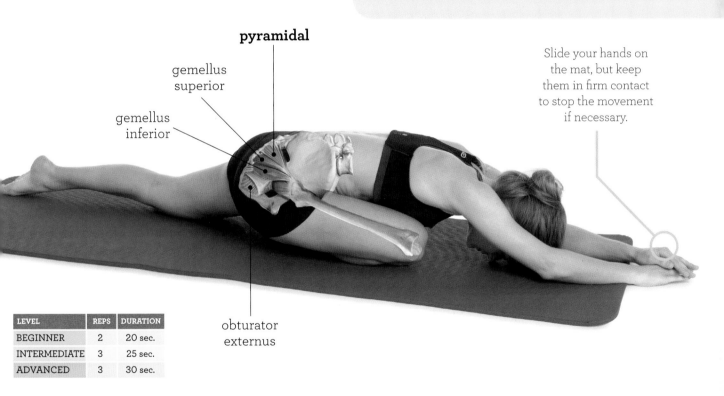

pyramidal

gemellus superior

gemellus inferior

Slide your hands on the mat, but keep them in firm contact to stop the movement if necessary.

obturator externus

LEVEL	REPS	DURATION
BEGINNER	2	20 sec.
INTERMEDIATE	3	25 sec.
ADVANCED	3	30 sec.

CAUTION

Do the movement slowly without letting yourself fall suddenly, because you probably will not achieve the degree of incline shown by the model in the photo.

BENEFITS

Relief of tension in the muscles in the rear of the hip and increased range of movement in the hip.

INDICATION

For athletes, especially those whose sport is demanding on the lower body or who experience muscle tightness and stiffness in the muscles in this area.

LEG AND FOOT STRETCHES

QUADRICEPS FEMORIS
This powerful, large muscle is comprised of four muscles of lesser size whose main common function is straightening the knee.

Rectus femoris: This muscle originates at the anteroinferior iliac spine, and it inserts at the quadricipital tendon that goes to the kneecap, and from there it is referred to as the patellar tendon, and it continues to its insertion point at the anterior tuberosity of the tibia.

Crural: This muscle originates at the proximal epiphysis of the femur, and it shares an insertion with the rectus femoris and the vastus medialis and lateralis.

Vastus lateralis and vastus medialis: These muscles originate at both sides of the epiphysis and the proximal third of the femur.

ISCHIOTIBIALS
These are three muscles that are found in the rear of the thigh. Their main function is bending the knee.

Biceps femoris: This muscle originates at the ischium and at the diaphysis of the femur, and it inserts at the proximal epiphysis of the tibia and the fibula.

Semitendinosus: This muscle originates at the ischium, and inserts at the proximal third of the diaphysis of the tibia.

Semimembranosus: This muscle originates at the ischium, and inserts at the proximal epiphysis of the tibia.

GASTROCNEMIUS
Known popularly as the calves, these muscles are made up of medial and lateral parts. They originate at the sides of the distal epiphysis of the femur, and they insert at the rear of the heel bone via the Achilles tendon. Their function is plantar flexion of the ankle.

SOLEUS
This muscle originates at the proximal epiphysis of the fibula and the diaphysis of the tibia and the fibula, and it shares an insertion with the gastrocnemius at the rear of the heel bone via the Achilles tendon.

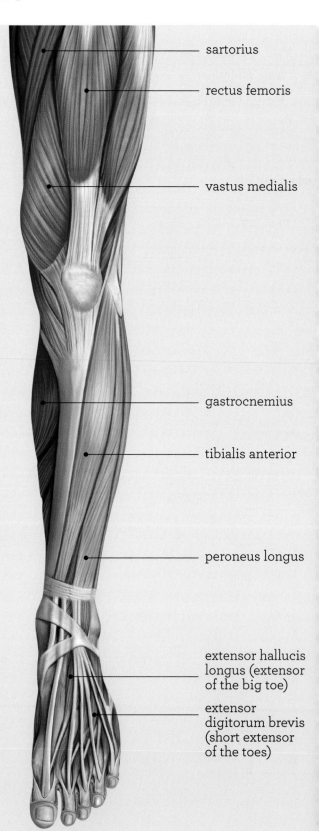

sartorius

rectus femoris

vastus medialis

gastrocnemius

tibialis anterior

peroneus longus

extensor hallucis longus (extensor of the big toe)

extensor digitorum brevis (short extensor of the toes)

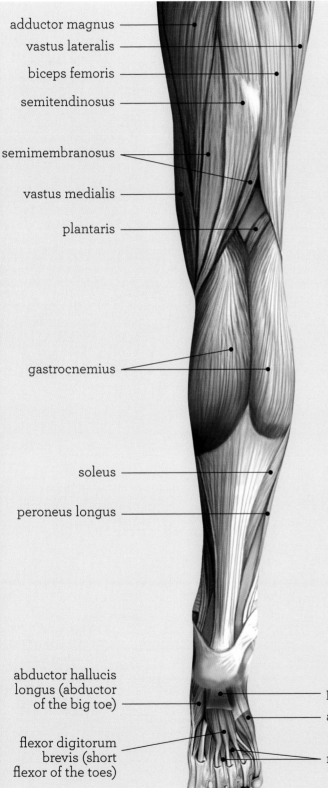

adductor magnus

vastus lateralis

biceps femoris

semitendinosus

semimembranosus

vastus medialis

plantaris

gastrocnemius

soleus

peroneus longus

abductor hallucis longus (abductor of the big toe)

flexor digitorum brevis (short flexor of the toes)

plantar fascia

abductor digiti minimi pedis (abductor of the little toe)

musculi lumbricales pedis (lumbrical muscles of the foot)

TIBIALIS ANTERIOR

This muscle originates at the proximal two thirds of the tibia and at the interosseous membrane, and it inserts at the first cuneiform bone and the first metatarsal. Its main function is dorsal flexion of the ankle.

PERONEUS MUSCLES

These are three muscles located on the outside of the leg. **Peroneus brevis:** This muscle originates at the distal half of the fibula, and it inserts at the fifth metatarsal. Its main functions are eversion and plantar flexion of the ankle. **Peroneus longus:** This muscle originates at the head and the proximal half of the fibula, and it inserts at the first cuneiform bone and the first metatarsal. Its main functions are eversion and plantar flexion of the ankle. **Peroneus anterior:** This muscle originates at the distal third of the fibula and the interosseous membrane, and it inserts at the fifth metatarsal. Its main functions are eversion and dorsal flexion of the ankle.

PLANTAR FASCIA

This is a fibrous, tough membrane made up of connective tissue. It is located on the sole of the foot. It is triangular in shape, and it inserts at the lower face of the heel bone and at the first phalanges. It is responsible for maintaining the plantar arch, and it serves as the insertion for many foot muscles.

LUMBRICALS

These muscles originate at the tendons of the long common flexor of the toes, and they insert at the phalanges and the dorsal expansions of the tendons of the common extensor of toes 2 through 5. Their main function is flexion of those toes.

FLEXOR HALLUCIS BREVIS (short flexor of the big toe)

This muscle originates at the cuboid and lateral cuneiform bones, and it has a double insertion at the proximal phalange of the big toe. Its main function is flexion of the big toe.

Standing with Rear Foot Hold

START

Stand next to an object you can use to help you keep your balance. Bend one knee by raising your foot to the rear and close to your gluteal. Hold your ankle using the hand on the same side. Keep your back perpendicular to the floor.

TECHNIQUE

Pull your ankle upward so that your knee reaches its maximum degree of bend and your heel is even closer to your gluteal. You can increase the degree of stretch by moving the thigh of the bent leg slightly rearward with respect to the line of your upper body.

Starting Position

Extend your hip and move your thigh slightly rearward.

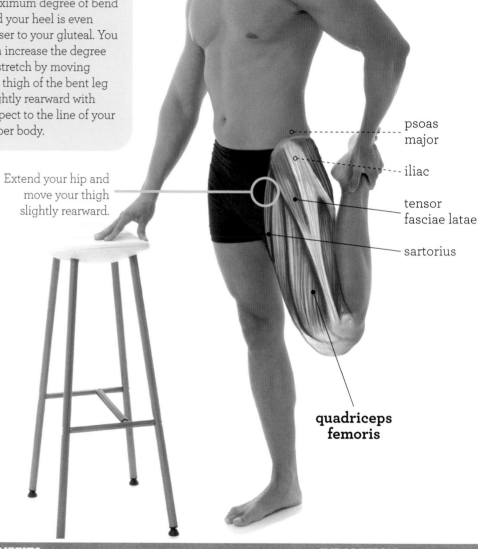

psoas major

iliac

tensor fasciae latae

sartorius

quadriceps femoris

LEVEL	REPS	DURATION
BEGINNER	2	25 sec.
INTERMEDIATE	3	35 sec.
ADVANCED	4	50 sec.

CAUTION

Hold onto a support point with your free hand to avoid losing your balance during this exercise.

BENEFITS

Broadening the range of movement in the knee and hip and reducing tension in the muscles in the front of the thigh.

INDICATION

For people who take part in regular physical activity or who suffer from hip and knee disorders, such as chondropathy or tendonitis.

Knight's Position with Rear Foot Hold

Starting Position

LEVEL	REPS	DURATION
BEGINNER	2	25 sec.
INTERMEDIATE	3	35 sec.
ADVANCED	4	50 sec.

START

Take a position on a mat and rest one knee on the floor in the knight's position (as shown in the illustration). If possible, use a support to keep your balance during the execution of this exercise.

TECHNIQUE

Hold the ankle of the rear leg using the hand on the same side, and pull upward while keeping your upper body perpendicular to the floor. You will feel the stretch in the front of your thigh. If the degree of stretch seems inadequate, you can lean forward slightly with your upper body without moving your support points, so that the hip corresponding to the leg being stretched reaches a greater degree of extension, and there is greater tension in the quadriceps.

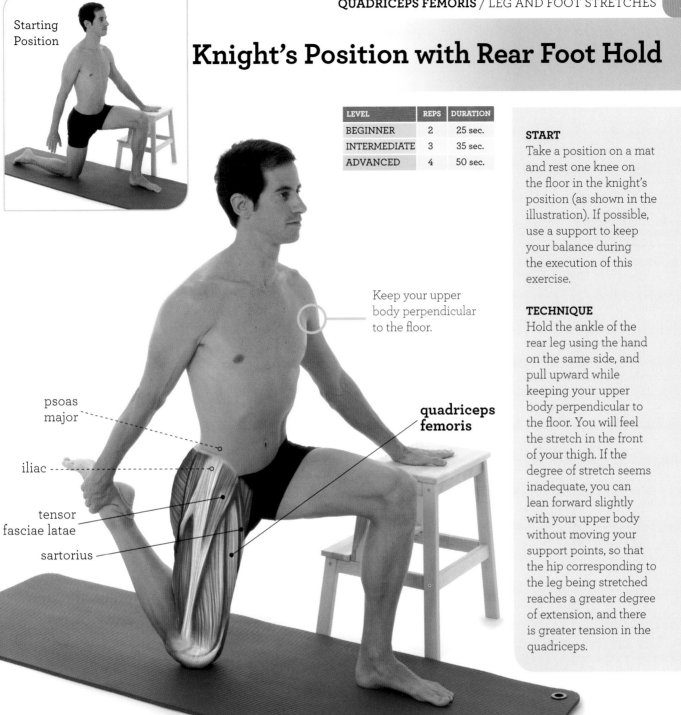

Keep your upper body perpendicular to the floor.

psoas major

iliac

tensor fasciae latae

sartorius

quadriceps femoris

CAUTION

Use your free hand to hold onto a support to keep your balance during this exercise.

BENEFITS

Broadening the range of movement in the knee and hip and reducing tension in the muscles in the front of the thigh.

INDICATION

For people who suffer from certain hip and knee disorders, such as chondropathy or tendonitis, and for athletes in general.

Bilateral on Knees

START
Kneel down on a mat. Keep your arms straight and perpendicular to the floor as you support yourself on both hands.

TECHNIQUE
By bending your knees, lower your upper body slowly and gradually until you feel the tension in the front part of your thighs. Even though you may go so far as to contact the floor with your elbows, this is not necessary if you produce enough stretch before reaching this point.

Starting Position

LEVEL	REPS	DURATION
BEGINNER	2	20 sec.
INTERMEDIATE	3	30 sec.
ADVANCED	4	35 sec.

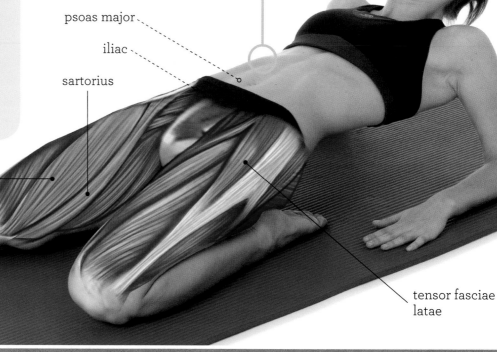

Keep your upper body aligned with your thighs.

psoas major

iliac

sartorius

quadriceps femoris

tensor fasciae latae

CAUTION
Maintain good support with your hands and lower yourself very gradually. If you feel muscle or knee pain, stop the descent immediately.

BENEFITS
Broadening of range of hip and knee movement and reducing tension in the muscles in the front of your thigh.

INDICATION
For athletes, regardless of their level.

Face-up Stretch

Starting
Position

LEVEL	REPS	DURATION
BEGINNER	2	20 sec.
INTERMEDIATE	3	30 sec.
ADVANCED	4	35 sec.

START
Sit down on a mat with your legs straight ahead and bend the knee of the leg to be stretched so that your thigh and calf contact one another.

TECHNIQUE
Slowly lean rearward with your upper body until you are lying on your back on the mat. You will feel the stretch in the front of your thigh as your upper body approaches the floor.

tensor fasciae
latae

psoas
major

sartorius

**quadriceps
femoris**

Keep your knee
totally bent.

iliac

CAUTION	BENEFITS	INDICATION
If you feel adequate tension in the front of your thigh before reaching the floor, stop the movement and hold the position. Every individual is different and may reach adequate degrees of stretch at different points in the range of movement. You may experience an injury if you push the stretch farther than you need to.	Broadening of the range of hip and knee movement and reducing tension in the muscles in the front of the thigh.	For people who take part in regular physical activity or who suffer from certain hip and knee disorders, such as chondropathy or tendonitis.

Stretch on Side

START
Lie down on one side, bend the knee of your uppermost leg, and hold your foot using the hand on the same side. The bottom leg and arm are aligned with your upper body.

TECHNIQUE
Pull on the foot you are grasping so that your knee reaches its maximum bend and your hip is extended. As you pull, the tension in the front part of your thigh will increase and you will stretch your quadriceps femoris.

Starting Position

LEVEL	REPS	DURATION
BEGINNER	2	25 sec.
INTERMEDIATE	3	35 sec.
ADVANCED	4	50 sec.

Pull on your foot to produce greater bend in your knee and greater hip extension.

quadriceps femoris

tensor fasciae latae

sartorius

iliac

psoas major

CAUTION	BENEFITS	INDICATION
You do this stretch in the usual way, with slow, controlled movements, because it presents no significant risk.	Broadening the range of movement in the knee and hip and reducing tension in the muscles in the front of the thigh.	For athletes and for people with chondropathy or tendonitis in the knee.

Standing with One Leg Forward

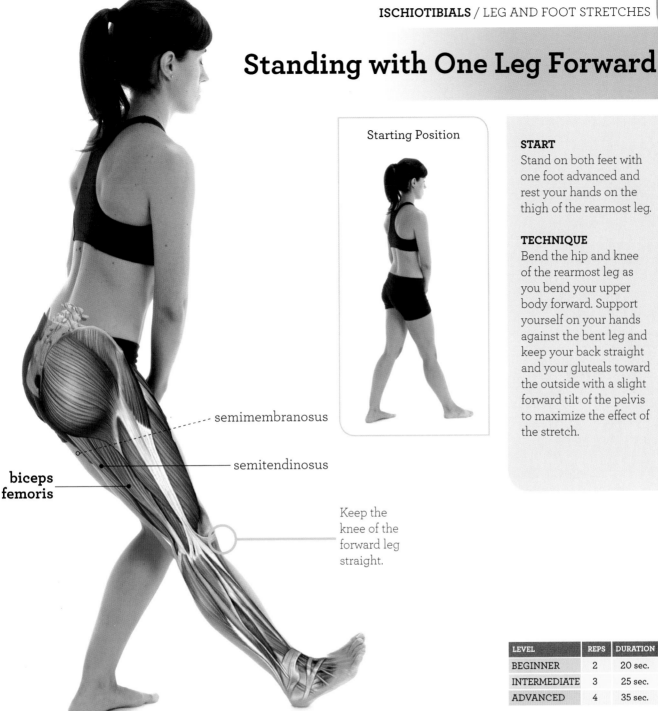

Starting Position

biceps
femoris

semimembranosus

semitendinosus

Keep the
knee of the
forward leg
straight.

START
Stand on both feet with
one foot advanced and
rest your hands on the
thigh of the rearmost leg.

TECHNIQUE
Bend the hip and knee
of the rearmost leg as
you bend your upper
body forward. Support
yourself on your hands
against the bent leg and
keep your back straight
and your gluteals toward
the outside with a slight
forward tilt of the pelvis
to maximize the effect of
the stretch.

LEVEL	REPS	DURATION
BEGINNER	2	20 sec.
INTERMEDIATE	3	25 sec.
ADVANCED	4	35 sec.

CAUTION

Do this exercise slowly and
progressively. Remember
that a little discomfort is
necessary for improvement,
but pain is the signal that
you are forcing the muscle
too much.

BENEFITS

Reduction of limitations in
hip and knee movement
and tension in the rear of
the thigh and lumbar area.
Contributes to maintenance
of correct posture.

INDICATION

For people with retroversion of
the pelvis and disappearance
of the lumbar curvature due
to tightness of the ischiotibial
muscles. Also for athletes.

Standing with Raised Leg

START
Stand next to a raised surface. You can use a chair, a stool, or anything else on which you can place your foot.

TECHNIQUE
Place your hands on the raised leg and bend your upper body forward. Slide your hands along your leg toward your ankle. Go slowly until you feel the tension in the back of your thigh.

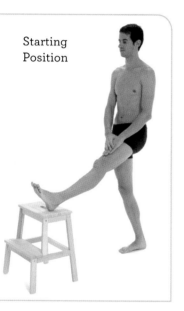

Starting Position

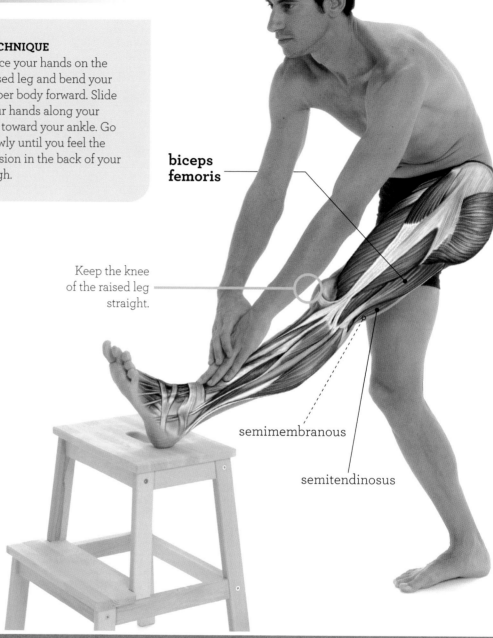

biceps femoris

Keep the knee of the raised leg straight.

semimembranous

semitendinosus

LEVEL	REPS	DURATION
BEGINNER	2	20 sec.
INTERMEDIATE	3	25 sec.
ADVANCED	4	35 sec.

CAUTION
Remember to avoid arching your back as you bend down in order to create less tension in your lumbar area and to maximize the effectiveness of the stretch.

BENEFITS
Maintaining proper range of movement in the lower body and relieving tension in the back of your thigh and lumbar area. It also contributes to maintaining correct posture.

INDICATION
For people with tight ischiotibials or posture problems due to retroversion of the pelvis and for athletes, especially those who require lower body power.

Knee Rest

Starting Position

START
Kneel on a mat and extend the leg to be stretched in front of you. Put one hand on the thigh of the forward leg as you bend your upper body to the front.

TECHNIQUE
Slide the hand resting on your thigh forward and try to reach the tip of your toes while keeping the knee of the forward leg straight. The tension in the back of your thigh and knee will indicate that the stretch is taking place.

LEVEL	REPS	DURATION
BEGINNER	2	20 sec.
INTERMEDIATE	3	25 sec.
ADVANCED	4	35 sec.

biceps femoris

Keep the knee of the forward leg totally straight.

semimembranosus

semitendinosus

CAUTION
If this position is unstable, rest your free hand on the floor.

BENEFITS
Maintaining normal functioning of the lower body by improving the range of movement, especially hip flexion and knee extension.

INDICATION
For people with tight ischiotibials, hip retroversion, and for athletes.

Bilateral Inverted V-position

START
Stand with your feet greater than shoulder-width apart and your back straight.

TECHNIQUE
Bend your upper body forward and straighten your arms as you try to touch the floor with your fingers. If this is too easy, you can reduce the distance between your feet and repeat the exercise, or else you can try to touch the floor with your knuckles or even the palms of your hands.

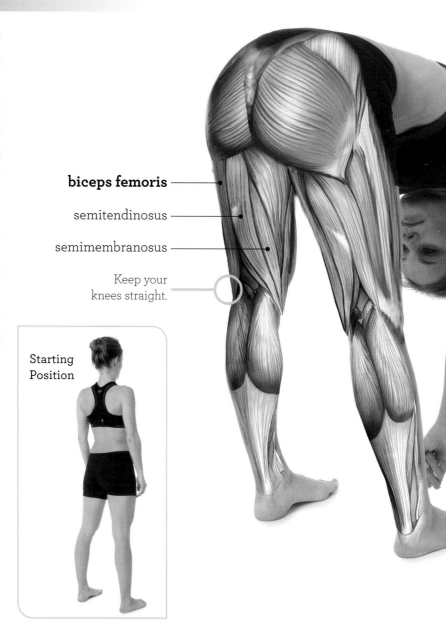

biceps femoris

semitendinosus

semimembranosus

Keep your knees straight.

Starting Position

LEVEL	REPS	DURATION
BEGINNER	2	20 sec.
INTERMEDIATE	3	25 sec.
ADVANCED	4	35 sec.

CAUTION	BENEFITS	INDICATION
Start with a balanced position and do the movement slowly, avoiding sudden jerks and inertia that could cause injury.	Increase in range of movement in straightening the knee and bending the hip and reduction of tension in this muscle group.	For people who experience tension in the back of the leg or lumbar problems due to tightness in the ischiotibials. Also for athletes, especially for those who require lower body power.

Bilateral Seated

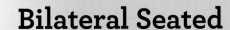

Starting Position

LEVEL	REPS	DURATION
BEGINNER	2	20 sec.
INTERMEDIATE	3	25 sec.
ADVANCED	4	35 sec.

START

Sit on a mat with your legs together and straight. Keep your back straight and perpendicular to the floor. Place your hands on your thighs and look straight ahead.

TECHNIQUE

Lean forward as you slide your hands toward your feet. As you move forward, you will feel the tension in the back of your thighs and knees.

semitendinosus

semimembranosus

biceps femoris

Keep your knees straight.

CAUTION	BENEFITS	INDICATION
Keep your knees straight to produce the stretch in your ischiotibials. Even a slight bend in the knees will detract from the exercise and reduce its effectiveness.	Increase in the range of movement in extending the knees and bending the hip and reduction in tension in the back of the thigh.	For people who experience tension in the back of the leg or lumbar problems due to tight ischiotibials. Also for athletes.

Supine with Raised Leg

START

Lie down on your back on a mat with your knees bent around 90°. Raise one leg toward your chest and hold onto it with both hands. Keep your head on the floor to avoid tension in your cervical vertebrae.

TECHNIQUE

Straighten the knee of the raised leg as you pull it toward your chest with your hands. The tension in the back of your thigh and knee will be the best indicator that you are doing the stretch correctly.

LEVEL	REPS	DURATION
BEGINNER	2	20 sec.
INTERMEDIATE	3	25 sec.
ADVANCED	4	35 sec.

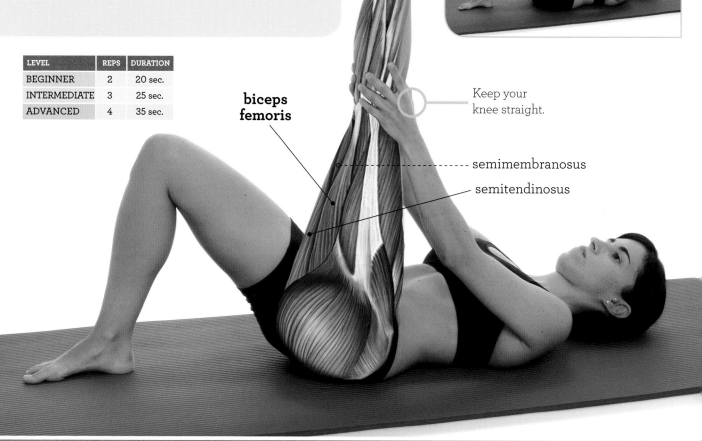

Starting Position

Keep your knee straight.

biceps femoris

semimembranosus

semitendinosus

CAUTION	BENEFITS	INDICATION
Do not go beyond the threshold of pain. A sense of discomfort is adequate for a proper stretch, and the ischiotibials can be injured by sudden jerks or excessive tension.	Increase in the range of movement in extending the knee or flexing the hip and reduction of tension in the back of the thigh.	For people with tension in the back of the thigh or hip retroversion due to tightness of the ischiotibials. Also for athletes, especially those who need lots of lower body strength.

Dorsiflexion Using Wall Support

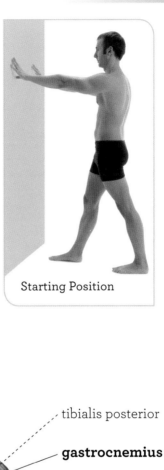

Starting Position

START
Stand in front of a wall at a distance that allows you to stretch your arms forward and touch the wall with your fingers. Both knees are straight, with one foot ahead of the other.

TECHNIQUE
Without moving your feet from their anchor points, move your upper body closer to the wall. You will have to bend the knee of the forward leg and your elbows. As you get closer to the wall, you will feel the tension in the calf of the rear leg. Hold the position for a few seconds and return to the starting point. Then do the stretch with the other leg.

tibialis posterior

gastrocnemius

Keep the heel of the rear leg on the floor.

soleus

plantaris

LEVEL	REPS	DURATION
BEGINNER	2	25 sec.
INTERMEDIATE	3	35 sec.
ADVANCED	4	40 sec.

CAUTION
Avoid bending the knee of the rear leg or lifting your heel, because the stretch would lose all its efficacy.

BENEFITS
Relief from tension in the calf and an increase in the range of movement in the ankle.

INDICATION
For people who commonly wear high heels, spend lots of time standing or walking, and athletes, especially swimmers, runners, and cyclists. Also for people with specific disorders of the ankle, tendonitis of the Achilles tendon, or pain in the sole of the foot.

Dorsiflexion Using Step

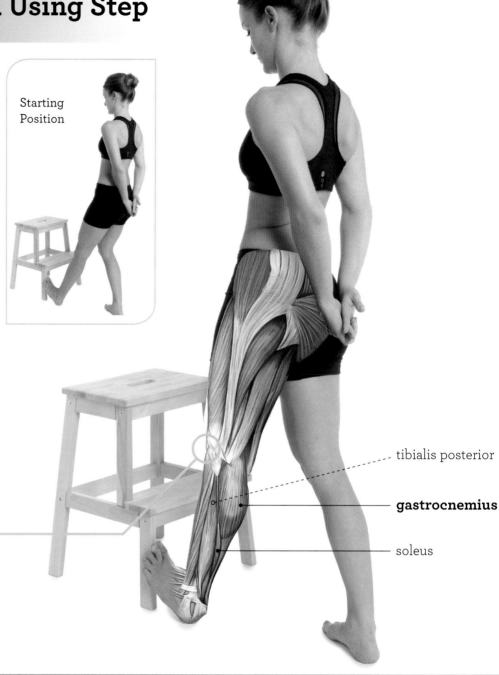

START

Stand in front of a step or other low object that you can use for support. Advance one foot, and rest the heel on the floor and your toes on the step. The rear foot acts as the main support, and your knee must be slightly bent, which will allow you to lean slightly rearward.

TECHNIQUE

Straighten the knee of the rear leg and move your body slightly forward without moving your feet from their anchor points. This will produce dorsiflexion of the ankle joint and tension in your calf due to the stretch of the gastrocnemius.

Starting Position

Keep the knee of the forward leg straight.

tibialis posterior

gastrocnemius

soleus

LEVEL	REPS	DURATION
BEGINNER	2	25 sec.
INTERMEDIATE	3	35 sec.
ADVANCED	4	40 sec.

CAUTION

Make sure that the front foot is supported on a fixed point so it doesn't move from the thrust produced by the exercise.

BENEFITS

Relaxation of the muscles in the back of the leg and maintenance of a full range of movement in the ankle joint.

INDICATION

For athletes, especially those who swim, run, or cycle. Also for people who wear high-heeled shoes or spend lots of time standing or walking, who experience strain or cramps in their calves or the soles of their feet, and who have specific ankle disorders, tendonitis of the Achilles tendon, or pain in the sole of the foot.

Seated Traction

Starting Position

Keep your knee straight.

soleus

plantaris

tibialis posterior

gastrocnemius

START
Sit on a mat with one leg drawn up and the other stretched out straight. The toes on the foot of the straight leg point rearward. Place your hands on the straight leg.

TECHNIQUE
Lean forward and hold the tip of the front foot with one hand. As you hold it, pull it rearward until you feel the tension from the stretch in the back of your leg. Hold the position for a few seconds and return to the starting point. Then perform the stretch with the other leg.

LEVEL	REPS	DURATION
BEGINNER	2	25 sec.
INTERMEDIATE	3	35 sec.
ADVANCED	4	40 sec.

CAUTION
Use the hand on the same side as the leg being stretched to pull back on the foot, because it has a greater margin of movement. If you use the opposite hand, you probably will feel tension in your ischiotibials before reaching your foot. If it is still difficult to reach your foot, you can use a towel.

BENEFITS
Relief from tension in the calves and improvement in the range of ankle movement.

INDICATION
For people who play sports, especially runners, swimmers, and cyclists. Also for those who stand or walk for prolonged periods, regularly wear high heels, or suffer from specific ankle disorders, tendonitis of the Achilles tendon, or pain in the sole of the foot.

Front Support

START

Stand in front of an object you can use for support, such as a tall stool, a chair, or something similar. Advance one foot and rest the tip of it on the furniture so that your heel is resting on the floor and your knee is straight. Your upper body will lean slightly forward, with your hands holding the support.

TECHNIQUE

Move your upper body forward as you bend both knees slightly and come closer to the support. The ankle dorsiflexion will increase, and you will feel the tension from the stretch in the back of your leg.

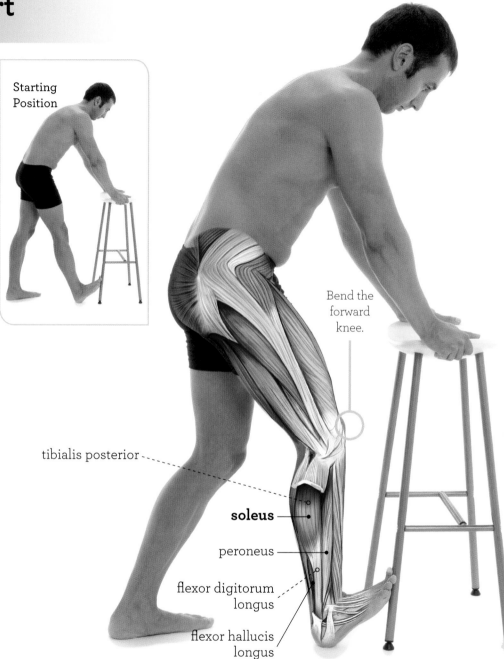

Starting Position

Bend the forward knee.

tibialis posterior

soleus

peroneus

flexor digitorum longus

flexor hallucis longus

LEVEL	REPS	DURATION
BEGINNER	2	25 sec.
INTERMEDIATE	3	35 sec.
ADVANCED	4	45 sec.

CAUTION

Rest the front of the forward foot against a stable object that will not move as a result of the thrust of the exercise.

BENEFITS

Relaxation of the muscles in the back of the leg and optimization of the range of ankle movement.

INDICATION

For people who spend lots of time standing or walking and for athletes, especially runners, cyclists, and swimmers. Also for people with specific ankle disorders, tendonitis of the Achilles tendon, pain in the sole of the foot, or who regularly wear high heels.

Bilateral on Step

Starting Position

START
Take a position on a step, in which you contact the step only with the front of your feet. This exercise produces a very unstable position, so it is recommended that you use some kind of handhold, a railing or other similar object, so you don't lose your balance while doing the exercise.

TECHNIQUE
Perform ankle dorsiflexion such that your heels are lower than the front of your feet and your entire body lowers slightly.

Keep both knees slightly bent.

tibialis posterior

soleus

flexor hallucis longus (flexor of the big toe)
flexor digitorum longus (long flexor of toes)
peroneus

LEVEL	REPS	DURATION
BEGINNER	2	20 sec.
INTERMEDIATE	3	25 sec.
ADVANCED	3	30 sec.

CAUTION
Make sure the front of your feet are secure on the step, and, if possible, hold onto a support.

BENEFITS
Improvement in range of ankle movement and relief from tension in the calves.

INDICATION
For runners and cyclists, people who spend lots of time standing or walking, people who commonly wear high heels, and people with specific problems of the ankle or Achilles tendon or pain in the sole of the foot.

Seated Traction with Both Hands

START

Sit on a mat with one leg straight and the other one drawn up. Place both hands near the ankle of the drawn-up leg.

TECHNIQUE

Grasp the foot of the drawn-up leg and pull on it while keeping your heel on the floor. This traction will produce tension in your calf by stretching the soleus.

LEVEL	REPS	DURATION
BEGINNER	2	25 sec.
INTERMEDIATE	3	30 sec.
ADVANCED	4	40 sec.

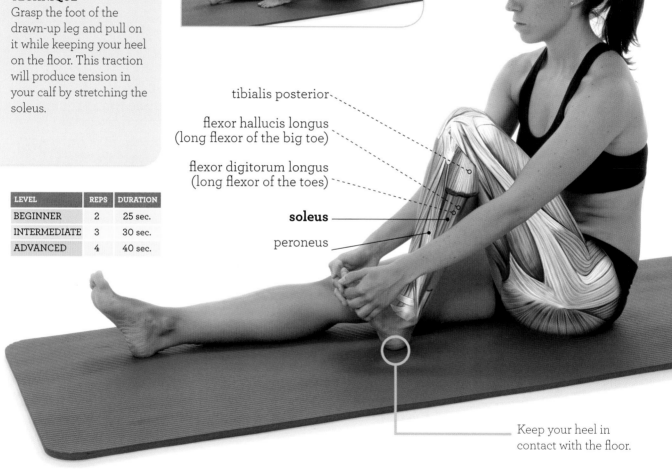

Starting Position

tibialis posterior

flexor hallucis longus (long flexor of the big toe)

flexor digitorum longus (long flexor of the toes)

soleus

peroneus

Keep your heel in contact with the floor.

CAUTION

You should not pull on your toes, but at the level of the metatarsophalangeal joints.

BENEFITS

Relief from possible tension in the calf and maintenance of optimal range of movement in the ankle.

INDICATION

For people with ankle and Achilles tendon disorders, for athletes who run, bicycle, or swim, and for people who wear high heels or spend lots of time standing or walking.

Bilateral on All Fours

Starting Position

START
Get down on all fours on a mat with your hands shoulder-width apart and your knees in line with your hips. Your feet should contact the mat on tiptoes.

TECHNIQUE
Without changing any of your anchor points, move your body rearward so that the bend in your knees increases and you end up nearly sitting on your calves. You will feel the tension in the soles of your feet and the back of your legs.

soleus

Contact the mat with the tips of your toes.

flexor digitorum longus (long flexor of the toes)

tibialis posterior

flexor hallucis longus (long flexor of the big toe)

LEVEL	REPS	DURATION
BEGINNER	2	20 sec.
INTERMEDIATE	3	25 sec.
ADVANCED	4	35 sec.

CAUTION
In this instance, contact with the toes is unavoidable, so the flexor muscles of the toes will also be stretched, and you will need to be careful with the tension you apply.

BENEFITS
Relief from tension in the calf and the sole of the foot and maintenance of an optimal range of movement in the ankle.

INDICATION
For swimmers, runners, and people who spend lots of time standing, walking, or wearing high heels. Also for those who suffer from ankle or Achilles tendon disorders or cramps in the soles of the feet.

Bilateral with Support on Insteps

START

Kneel on a mat and bend your knees until you are sitting on your calves. Support yourself with your hands at your sides and slightly toward the front.

TECHNIQUE

Raise your knees from the floor while keeping your insteps in contact with the mat. Your arms will support part of your body weight. The plantar flexion of the ankle will produce tension in the front of your leg and will stretch the tibialis anterior.

Starting Position

extensor hallucis longus (extensor of the big toe)

Maintain the support on your insteps.

peroneus anterior

extensor digitorum communis

tibialis anterior

LEVEL	REPS	DURATION
BEGINNER	2	20 sec.
INTERMEDIATE	3	25 sec.
ADVANCED	4	35 sec.

CAUTION

Make sure that the surface beneath you is slightly padded so you don't hurt your ankles and insteps.

BENEFITS

Relief from tension in the front of the legs and maintenance of an optimal range of ankle movement.

INDICATION

For athletes, especially for skaters and skiers.

Posterior on Instep

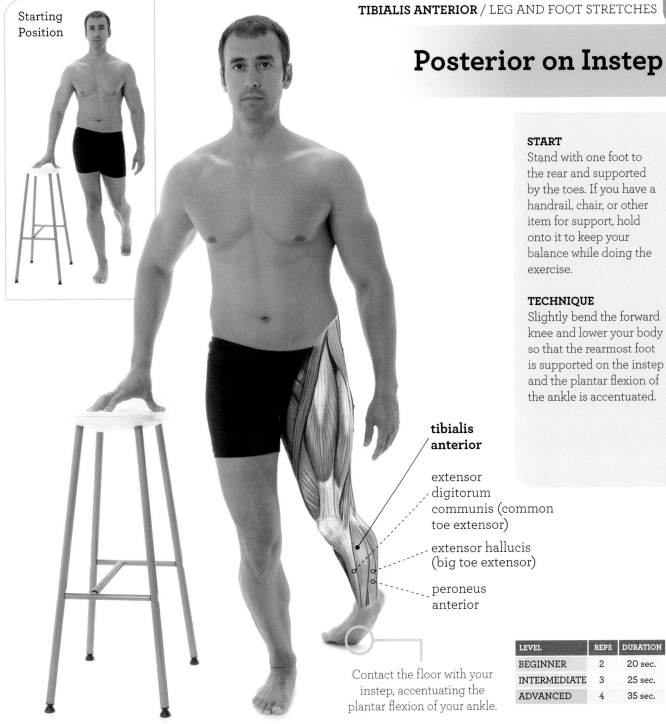

Starting Position

START
Stand with one foot to the rear and supported by the toes. If you have a handrail, chair, or other item for support, hold onto it to keep your balance while doing the exercise.

TECHNIQUE
Slightly bend the forward knee and lower your body so that the rearmost foot is supported on the instep and the plantar flexion of the ankle is accentuated.

tibialis anterior

extensor digitorum communis (common toe extensor)

extensor hallucis (big toe extensor)

peroneus anterior

Contact the floor with your instep, accentuating the plantar flexion of your ankle.

LEVEL	REPS	DURATION
BEGINNER	2	20 sec.
INTERMEDIATE	3	25 sec.
ADVANCED	4	35 sec.

CAUTION	**BENEFITS**	**INDICATION**
Use some type of support for stability during the stretch.	Relaxation of the muscles in the front of the leg and maintenance of an optimal range of ankle movement.	For athletes, especially those who use skates or skis in their sports.

Seated Lateral Pull

START

Sit on a mat with one leg drawn up and the other one stretched out straight. Lean forward and contact your foot with the opposite hand, so that your fingers are on your instep.

TECHNIQUE

Pull back on your foot to generate tension in the peroneus. Hold the stretch for a few seconds before returning to the starting point.

LEVEL	REPS	DURATION
BEGINNER	2	20 sec.
INTERMEDIATE	3	25 sec.
ADVANCED	4	35 sec.

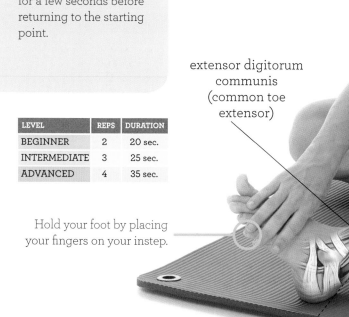

Starting Position

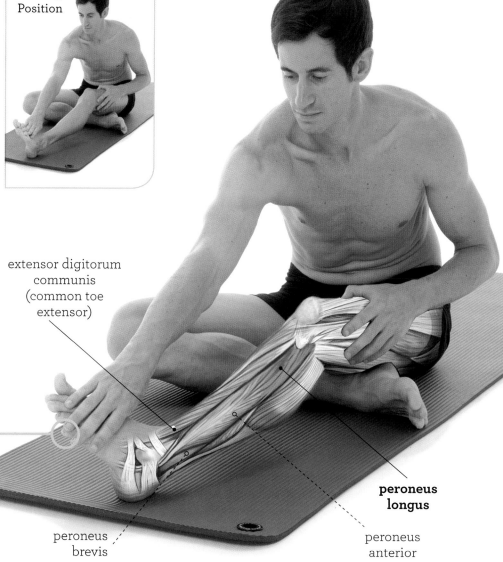

extensor digitorum communis (common toe extensor)

Hold your foot by placing your fingers on your instep.

peroneus brevis

peroneus longus

peroneus anterior

CAUTION	**BENEFITS**	**INDICATION**
Avoid forcing the lumbar region as you bend forward. If you feel any pain, choose some other exercise for the same muscle group.	Relief from tension in the outside of the leg and maintenance of a broad range of movement in the ankle.	For athletes and people who have experienced an ankle injury, because the peroneus muscles are subject to strain after this type of injury.

Bilateral on Knees

Starting Position

START
Get down on all fours on a mat so that your contact points are both hands and your toes. Your thighs and calves must be in close contact with one another, and you are practically sitting on your heels.

TECHNIQUE
Lower your knees until they contact the floor, so that the extension of your toes is accentuated and the tension is increased in the flexor muscles of the toes.

flexor digitorum brevis (short toe flexor)

flexor hallucis brevis (short flexor of the big toe)

lumbricals

Contact the mat with your toes in extension.

LEVEL	REPS	DURATION
BEGINNER	2	20 sec.
INTERMEDIATE	3	25 sec.
ADVANCED	3	35 sec.

CAUTION
If you feel pain in your toes at any point in the movement, stop without going all the way.

BENEFITS
Relaxation of the muscles in the sole of the foot and the calf.

INDICATION
For people who experience strain or cramps in the sole of the foot and athletes, especially swimmers and those who take part in sports in which running is important.

Seated Foot Pull

START
Sit with one leg straight and the other one drawn up. Hold the foot of the drawn-up leg with both hands. Grasp your toes with one hand and your heel with the opposite hand.

TECHNIQUE
Pull rearward on your toes to produce the maximum extension while keeping your heel stationary.

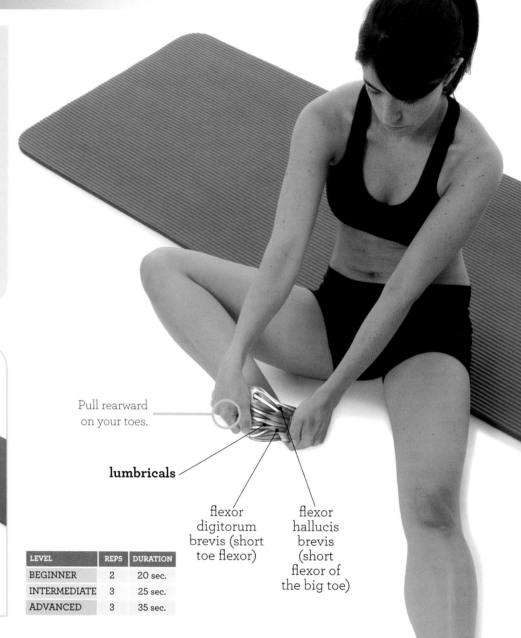

Pull rearward on your toes.

lumbricals

flexor digitorum brevis (short toe flexor)

flexor hallucis brevis (short flexor of the big toe)

Starting Position

LEVEL	REPS	DURATION
BEGINNER	2	20 sec.
INTERMEDIATE	3	25 sec.
ADVANCED	3	35 sec.

CAUTION
Remember that you are exerting tension on small body parts, so the pull must be moderate and controlled at all times.

BENEFITS
Relief from tension in the sole of the foot.

INDICATION
For athletes, especially those who engage in sports that involve swimming. Also for people who experience cramps in the sole of the foot.

Toe Pull

Starting Position

START
Sit on a mat with both legs bent in front or crossed (as shown in the illustration). Keep the sole of the foot to be stretched in contact with the floor and hold the toes of this foot with your hand.

TECHNIQUE
Using the hand holding your toes, pull back gently on them, bringing them to maximum extension. Hold this position for a few seconds while maintaining contact between the sole of your foot and the floor.

Pull back on your toes without lifting the sole of your foot.

flexor hallucis brevis (short flexor of the big toe)

lumbricals

flexor digitorum brevis (short flexor of the toes)

LEVEL	REPS	DURATION
BEGINNER	2	20 sec.
INTERMEDIATE	3	25 sec.
ADVANCED	3	35 sec.

CAUTION
Avoid reaching a point where you feel pain, because small joints, along with other particularly unstable ones, such as the shoulder, are susceptible to injury when subjected to extreme traction.

BENEFITS
Relief from tension in the sole of the foot and the toes, reduction in the possibility of experiencing pain or cramps and relief as soon as they occur.

INDICATION
For people who swim or do other sports connected with water, such as water polo, and scuba diving. Also for people who regularly participate in dancing.

STRETCHING GUIDE
FOR AILMENTS

- Spinal Pain
- Back Pain
- Lower Back Pain
- Shoulder Pain
- Elbow Pain
- Hand Pain
- Pelvis and Hip Pain
- Gluteal Pain
- Knee Pain
- Leg Pain
- Ankle Pain
- Sole of the Foot Pain

NECK PAIN

The pain that is felt in the neck area is known as *cervicalgia*. This is not a specific ailment, but rather a symptom of various disorders that may arise in the cervical area. Cervicalgia may occur as a result of poor posture, spending lots of time behind the steering wheel or at the computer, or cervical whiplash (as in rear-end automobile collisions). In most instances, the pain is localized at the rear and side of the neck, and it originates in the muscles and joints.

1 p. 30 LATERAL NECK BEND
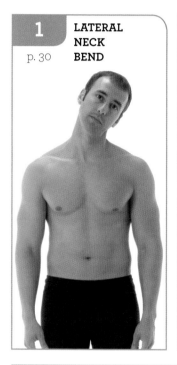

2 p. 31 ASSISTED NECK BEND
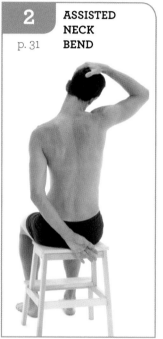

3 p. 32 NECK BEND AND ROTATION

6 p. 35 ASSISTED NECK BEND

13 p. 44 ASSISTED UPPER BODY BEND
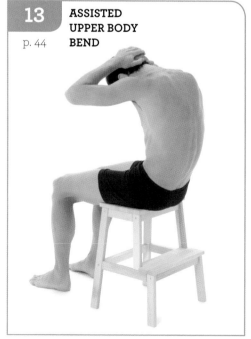

27 p. 60 CROUCHING UPPER BODY BEND

NECK/SHOULDER/ARM PAIN

The pain that appears in the cervical area, extends through the shoulder and arm, and sometimes goes as far as the hand or the chest is known as *cervicobrachialgia*. It often originates in the nerves, because of a pinching of a nerve root. This may be due to a muscle contraction. In these cases, the cervicobrachialgia may diminish or disappear with stretching.

| 1 | p. 30 | LATERAL NECK BEND |

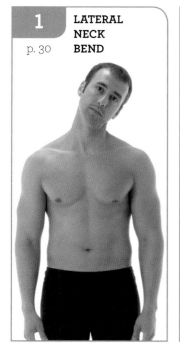

| 2 | p. 31 | ASSISTED NECK BEND |

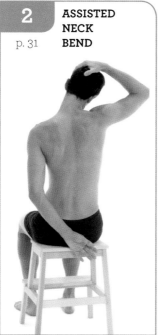

| 3 | p. 32 | NECK BEND AND ROTATION |

| 4 | p. 33 | NECK ROTATION AND EXTENSION |

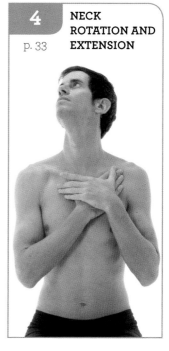

| 5 | p. 34 | NECK EXTENSION AND CHIN RAISE |

| 6 | p. 35 | ASSISTED NECK BEND |

BACK PAIN

Dorsalgia is pain that occurs in the dorsal region of the spinal column, between the cervical and lumbar areas. It may have various origins, such as posture problems, positions held for a long time while working, scoliosis, hyperkyphosis, muscle imbalances, trauma, or other organic causes. In cases where the dorsalgia is due to posture problems or muscle imbalances, it may be relieved with stretching.

9 p. 40 — **MOHAMMED POSITION**

16 p. 47 — **LEG HUG**

11 p. 42 — **RAISED ARM UPPER BODY BEND**

12 p. 43 — **TRACTION FROM FIXED POINT**

13 p. 44 — **ASSISTED UPPER BODY BEND**

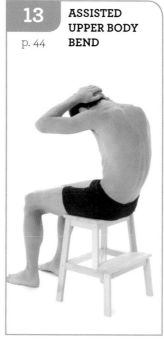

14 p. 45 — **ARMS FORWARD**

LOWER BACK PAIN

Lower back pain is the pain produced by bad posture, which itself is caused by anteversion or retroversion of the pelvis, improper lifting of heavy weights, maintaining harmful positions (generally at work), and, to a lesser degree, other, more serious problems, including a herniated disk or vertebral arthrosis or stenosis. In these latter cases, you need to get medical advice before starting any stretching program on your own.

The *acute phase* refers to the moment of maximum pain. It may last a few days, and the pain is very intense and limiting.

23 **STRETCH WITH KNEES AGAINST CHEST**
p. 56

25 **KNEE BEND ONTO CHEST**
p. 58

24 **CROSSED LEG**
p. 57

87 **SUPINE WITH RAISED LEG**
p. 134

LOWER BACK PAIN

Maintenance Phase: This is the phase in which the pain diminishes, appears briefly, or even disappears. In this phase, it is appropriate to do stretches to eliminate the pain and reduce the risk of recurrence.

23
p. 56
STRETCH WITH KNEES AGAINST CHEST

24
p. 57
CROSSED LEG

25
p. 58
KNEE BEND ONTO CHEST

68
p. 113
KNIGHT'S POSITION

26
p. 59
SEATED WITH UPPER BODY BEND

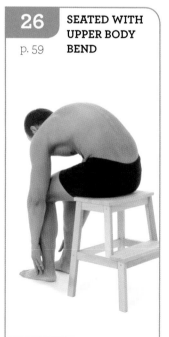

67
p. 112
UNILATERAL WITH STEP

69
p. 114
KNEE BEND AND OPPOSITE LEG EXTENSION

87
p. 134
SUPINE WITH RAISED LEG

SHOULDER PAIN

Generally speaking, pain in the shoulder joints originates in muscles, tendons, or the joint itself, and it may arise as the result of athletic participation, intense physical labor, or through forced movements.

Rotator Cuff Damage: The shoulder is a highly mobile joint, and this implies greater instability and risk of injury. One of the most common problems is damage to the rotator cuff, which provides the joint with stability. Generally, this damage is produced by inflammation of the tendon of one of the rotator muscles as a result of friction, overuse, or trauma, and this especially affects people who do activities in which the elbows are regularly raised higher than the shoulders.

2 ASSISTED NECK BEND
p. 31

29 POSTERIOR WITH ARM IN FRONT
p. 66

34 UNILATERAL WALL SUPPORT
p. 71

35 SUPPORT WITH BENT ELBOW
p. 72

39 FORWARD ELBOW PULL
p. 76

41 INVERTED UNILATERAL SUPPORT
p. 80

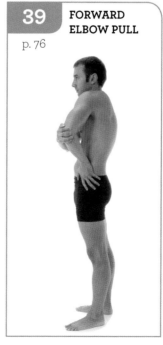

ELBOW PAIN

This involves pain in the tendons that generally result from intense athletic activity using repetitive movements.

Epicondylitis/Tennis Elbow: This is pain in the outer side of the elbow produced by repetitive wrist movements that affect the muscles and tendons used for extending the wrist and for supination of the forearm. It affects athletes in racquet sports, especially as a result of repeated backhand shots, but it also affects street sweepers, cleaners, illustrators, and people who do manual labor.

Epitrochleitis/Golfer's Elbow: This is pain on the inside of the elbow produced by repetitive flexion movements of the wrist and by pronation of the forearm, which cause small injuries and inflammation in the tendons of the relevant muscles. This disorder commonly affects golfers and people who do manual labor.

41 p. 80 — **INVERTED UNILATERAL SUPPORT**

42 p. 81 — **WALL SUPPORT WITH TWIST**

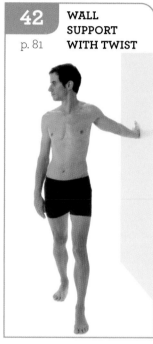

50 p. 89 — **BILATERAL WITH BACK HAND SUPPORT**

49 p. 88 — **BENT WRIST PULL**

51 p. 92 — **WRIST AND FINGER FLEX**

47 p. 86 — **WRIST PULL AND EXTENSION**

52 p. 93 — **ASSISTED WRIST EXTENSION**

HAND PAIN

This pain commonly affects people who perform repetitive tasks with their hands. Moving the hands and fingers in repeated, forced tasks can produce various problems.

Dupuytren's Contracture: This affects the palmar aponeurosis, causing it to shrink, so that the hand or the fingers gradually close. It is most common in people who regularly handle heavy weights.

De Quervain's Tenosynovitis: This affects the tendons of the short extensor and the long abductor of the thumb. It causes pain and functional loss, and it affects primarily people whose activity requires the use of hand tools, such as butchers, fishermen, carpenters, and bricklayers, and people who regularly participate in sports that require implements, such as rackets, sticks, and golf clubs. It also occurs in mothers of newborns, because of the need to handle the baby frequently in providing care.

55 RHOMBUS POSITION
p. 96

54 THUMB BEND
p. 95

52 ASSISTED WRIST EXTENSION
p. 93

53 FINGER EXTENSION
p. 94

49 BENT WRIST PULL
p. 88

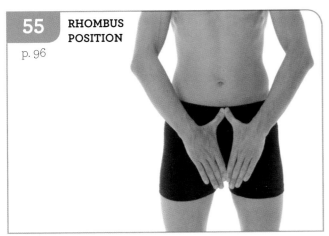

51 WRIST AND FINGER FLEX
p. 92

PELVIS AND HIP PAIN

Pain or discomfort in the pelvic area occurs most frequently when doing activities and participating in sports that involve running and jumping.

Dynamic Osteopathy of the Pubis/Pubalgia: This is pain in the area of the pubis that generally results from regular athletic practice that ends up causing problems in the tendons of the adductors and the abdominal muscles. Sometimes it can be due to trauma and bone ailments.

19 COBRA POSITION p. 52	20 ROTATION STRETCH p. 53

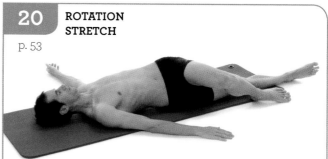

58 LEG EXTENSION ON ALL FOURS p. 103	61 BACKWARD MOVEMENT ON KNEES AND FOREARMS p. 106

60 BILATERAL IN SUMO POSITION p. 105	67 UNILATERAL WITH STEP p. 112	62 BUTTERFLY POSITION p. 107

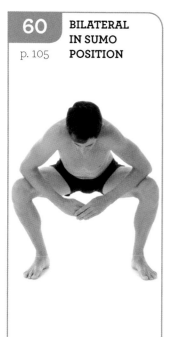

PELVIS AND HIP PAIN

Trochanteric Bursitis: This is an inflammation of the bursa that covers the trochanter major of the femur. It commonly results from repetitive friction with the iliotibial band in flexion and extension movements of the hip, so it especially affects women, because of their anatomical makeup

(outward hip rotation), and running and walking athletes, especially long-distance ones. It produces pain in the area of the proximal epiphysis of the femur.

Hip Impingement: This condition results when the head or the neck of the femur rubs against an edge of the acetabulum, because of irregular growth in one of these two parts. Generally, pain occurs when the irregular growth is very pronounced or when the hip joint is subjected to abrupt or extreme movements, such as those performed by gymnasts and athletes who practice martial arts or synchronized swimming.

65   **LATERAL UPPER BODY FLEX WITH CROSSED LEG**   p. 110

66   **UNILATERAL STANDING WITH SUPPORT**   p. 111

69   **KNEE BEND AND OPPOSITE LEG EXTENSION**   p. 114

71   **UNILATERAL WITH CROSSED LEG**   p. 116

67   **UNILATERAL WITH STEP**   p. 112

68   **KNIGHT'S POSITION**   p. 113

72   **SUPINE KNEE AND HIP FLEX**   p. 117

GLUTEAL PAIN

Pain that commonly occurs in athletes who participate in foot races and cycling, as well as in people who take part in team sports. This pain is commonly caused by muscle strain and small injuries to the muscles.

24 p. 57 — CROSSED LEG

69 p. 114 — KNEE BEND AND OPPOSITE LEG EXTENSION

70 p. 115 — KNEE BEND WITH OPPOSITE HEEL REARWARD

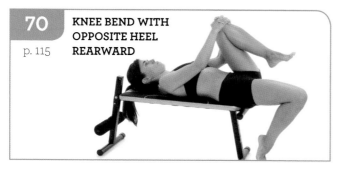

71 p. 116 — UNILATERAL WITH CROSSED LEG

72 p. 117 — SUPINE KNEE AND HIP FLEX

73 p. 118 — CROSS ON KNEE

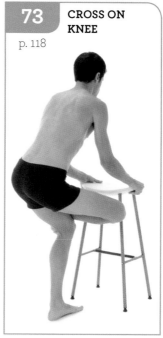

87 p. 134 — SUPINE WITH RAISED LEG

76 p. 121 — FACE-DOWN CROSS

KNEE PAIN

Iliotibial Band Disease: This involves inflammation of the iliotibial band or belt at the point where it passes next to the lateral epicondyle of the femur, and it is produced by repetitive friction between these two parts. This pain generally occurs in long-distance runners, but it is not exclusive to them. It is accompanied by pain in the outer part of the knee.

Tendonitis of the Kneecap/Quadriceps: This is an inflammation or minor injury to the kneecap or quadriceps tendon that generally results from strain and is accompanied by pain in the front part of the knee. This strain commonly occurs in athletes whose sports involve frequent jumps, such as the long jump, triple jump, obstacle races, and volleyball.

65 p. 110 — **LATERAL UPPER BODY FLEX WITH CROSSED LEG**

66 p. 111 — **UNILATERAL STANDING WITH SUPPORT**

70 p. 115 — **KNEE BEND WITH OPPOSITE HEEL REARWOOD**

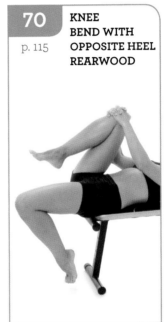

77 p. 124 — **STANDING WITH REAR FOOT HOLD**

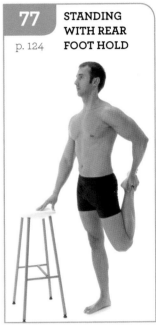

72 p. 117 — **SUPINE KNEE AND HIP FLEX**

78 p. 125 — **KNIGHT'S POSITION WITH REAR FOOT HOLD**

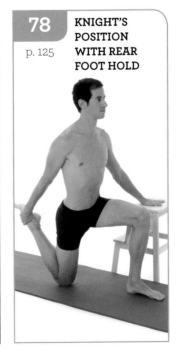

82 p. 129 — **STANDING WITH ONE LEG FORWARD**

97 p. 144 — **SEATED LATERAL PULL**

KNEE PAIN

Pes Anserinus or Goosefoot Syndrome: This is an inflammation of the goosefoot, which consists of the insertions of the tendons of the gracilis, sartorius, and semitendinosus muscles. Pain occurs on the inside of the knee. It is generally caused by strain, especially in running sports, due to excessive pronation of the foot and, more frequently, by trauma.

Patellar Chondropathy/Femoropatellar Syndrome: This is an injury to the patellar cartilage. It causes displacement of the kneecap on the femur during flexion and extension of the knee. Normally, it entails pain in the anterior region of the knee, and it is most common among athletes who subject their knees to great tension, such as runners, jumpers, and soccer and basketball players.

58 LEG EXTENSION ON ALL FOURS
p 103

70 KNEE BEND WITH OPPOSITE HEEL REARWARD
p. 115

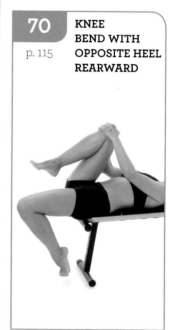

77 STANDING WITH REAR FOOT HOLD
p. 124

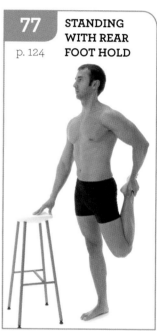

57 STANDING LEG EXTENSION
p. 102

60 BILATERAL IN SUMO POSITION
p. 105

82 STANDING WITH ONE LEG FORWARD
p. 129

87 SUPINE WITH RAISED LEG
p. 134

64 SUPINE WITH LEGS IN V-POSITION
p. 109

LEG PAIN

This type of problem is almost always due to the use of improper footwear or to regular participation in sports involving running. The result is excess tension in the tendons and muscles.

Tibial Periostitis: This is an inflammation of the tibial periosteum, the membrane that covers the bone. The condition causes pain in the anterior and inside area. It almost always is the result of exercise, and it affects foot racers, especially long- and middle-distance runners. It may arise through overtraining, a change in footwear, or the use of improper footwear.

91 FRONT SUPPORT
p. 138

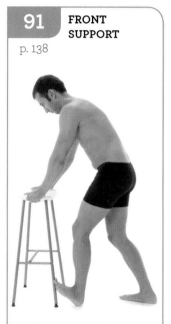

92 BILATERAL ON STEP
p. 139

93 SEATED TRACTION WITH BOTH HANDS
p. 140

94 BILATERAL ON ALL FOURS
p. 141

95 BILATERAL WITH SUPPORT ON INSTEPS
p. 142

96 POSTERIOR ON INSTEP
p. 143

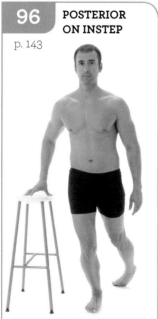

97 SEATED LATERAL PULL
p. 144

ANKLE PAIN

This is almost always due to improper footwear or to regular participation in sports involving running. It is caused by excessive tension in the tendons.

Tendonitis of the Achilles Tendon: The Achilles tendon connects the gastrocnemius and soleus muscles to the heel bone. This tendon can become inflamed or experience small injuries due to strain caused by running, especially frequent running on sand or irregular surfaces and ballet.

88 DORSIFLEXION USING WALL SUPPORT
p. 135

89 DORSIFLEXION USING STEP
p. 136

90 SEATED TRACTION
p. 137

93 SEATED TRACTION WITH BOTH HANDS
p. 140

91 FRONT SUPPORT
p. 138

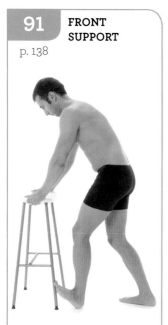

92 BILATERAL ON STEP
p. 139

94 BILATERAL ON ALL FOURS
p. 141

SOLE OF THE FOOT PAIN

Pain in the heel and sole of the foot is caused by strain in this area involving the plantar fascia.

Plantar fasciitis and heel spur: These two ailments commonly go together, because the occurence of one of them commonly causes or is caused by the other. Inflammation of the plantar fascia, which goes from the heel toward the front of the sole of the foot, causes pain and often occurs in long-distance runners and in people who run on uneven terrain or who use improper footwear.

89 DORSIFLEXION USING A STEP
p. 136

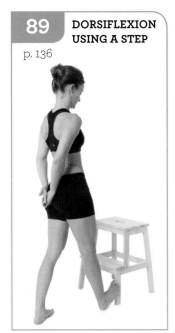

91 FRONT SUPPORT
p. 138

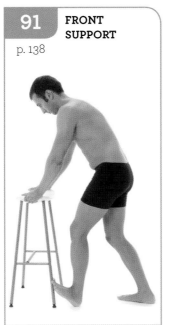

93 SEATED TRACTION WITH BOTH HANDS
p. 140

94 BILATERAL ON ALL FOURS
p. 141

99 SEATED FOOT PULL
p. 146

98 BILATERAL ON KNEES
p. 145

100 TOE PULL
p. 147

Alphabetical Index of Muscles

Musculi lumbricales manus, 90

Musculi lumbricales pedis, 123

Obliquus internus abdominus, 36

Obturator externus, 118, 119, 120, 121

Obturator internus, 100

Omohyoid, 29, 33, 34

Palmaris longus, 90

Palmaris minoris, 79

Pectinius, 102, 103, 104, 105, 106, 107, 108, 109, 114

Pectoral, 48

Pectoralis major, 64, 68, 69, 70, 71, 72, 73, 74, 77, 78, 81

Pectoralis minor, 64, 70, 71, 72, 74

Peroneus, 138, 140

Peroneus anterior, 142, 143, 144

Peroneus brevis, 144

Peroneus longus, 122, 123, 144

Plantar fascia, 123

Plantaris, 123, 135, 137

Posterior deltoid, 38, 39, 67

Pronator teres, 78

Psoas major, 101, 112, 113, 114, 115, 124, 125, 126, 127, 128

Psoas minor, 101

Pyramidal, 100, 118, 119, 120, 121

Quadratus femoris, 100, 124, 125, 126, 127, 128

Quadratus lumborum, 36, 49, 53, 55, 56, 57, 58, 59, 60, 61

Rectus abdominis, 48, 50, 51, 52

Rectus capitus anterior, 29

Rectus capitus lateralis, 29

Rectus capitus posterior major, 44

Rectus capitus posterior minor, 44

Rectus femoris, 122

Rhomboid, 47, 76

Rhomboideus, 37

Rhomboids, 28, 45, 46, 65, 75

Rotores lumborum, 36

Sartorius, 48, 112, 113, 115, 122, 124, 125, 126, 127, 128

Scalenes, 33, 34

Scalenus anterior, 29, 64

Scalenus posterior, 29

Scalenus medius, 29, 64

Semimembranosus, 129, 130, 131, 132, 133, 134

Semispinalis, 28, 32, 35, 36, 37, 44

Semitendinosus, 100, 123, 129, 130, 131, 132, 133, 134

Serratus anterior, 38, 39, 40, 41, 43, 48, 64, 74, 78

Serratus posterior inferior, 36

Serratus posterior superior, 36

Soleus, 123, 135, 136, 137, 138, 139, 140, 141

Spinal erectors, 61

Spinalis, 35

Splenius capitis, 28, 32, 35, 36, 37, 65

Splenius cervicis, 35

Sternocleidomastoid, 28, 29, 30, 31, 33, 34, 64

Subscapularis, 77

Supinator longus, 78, 81

Supraspinatus, 28, 37, 65

Tensor fasciae latae, 37, 48, 100, 110, 111, 116, 117, 124, 125, 126, 127, 128

Teres major, 39, 40, 41, 42, 43, 65, 70, 77, 83, 84, 85

Teres minor, 37, 38, 65, 66, 67, 75, 76

Tibialis anterior, 122, 142, 143

Tibialis posterior, 135, 136, 137, 138, 139, 140, 141

Transverses nuchae, 44

Transversospinalis, 36

Trapezius, 28, 29, 30, 31, 32, 35, 37, 45, 46, 47, 64, 65, 75

Triceps brachii, 37, 78, 79, 83, 84, 85

Vastus lateralis, 123

Vastus medialis, 122, 123

Bibliography

Alter, Michael, *Sport Stretch*, 2nd Edition. Champaign, Il: Human Kinetics, 1998.

Berg, Kristian, *Prescriptive Stretching*, Champaign, Il: Human Kinetics, 2011.

Blazevich, Anthony, *Sports Biomechanics: The Basics: Optimizing Human Performance*, 2nd Edition, London: Bloomsbury, 2010.

Delavier, Frederic, Clemenceau, Jean-Pierre, and Gundill, Michael, *Delavier's Stretching Anatomy*, Champaign, Il: Human Kinetics, 2011.

Frederick, Ann and Frederick, Christian, *Stretch to Win*, Champaign, Il: Human Kinetics, 2006.

Nelson, Arnold and Kokkonen, Jouko, *Stretching Anatomy*, 2nd Edition, Champaign, Il: Human Kinetics, 2014.

Norris, Christopher M., *The Complete Guide to Stretching*, 4th Edition, London: Bloomsbury, 2015.